Mathematics N4

E. Emede

GMT Publishers Phone: 065 895 1849 Fax: 011 604 2771 Email: gmtpublishers@gmail.com Tutorial Support: mathsmech206@gmail.com Illustrations: E. Emede Cover Design: E. Emede All rights reserved
ISBN 9798230985136

Mathematics N4

FET College Nated, Volume 6

Efetobo Emede

Published by Efetobo Emede, 2023.

While every precaution has been taken in the preparation of this book, the publisher assumes no responsibility for errors or omissions, or for damages resulting from the use of the information contained herein.

MATHEMATICS N4

First edition. December 27, 2023.

Copyright © 2023 Efetobo Emede.

ISBN: 979-8230985136

Written by Efetobo Emede.

Table of Contents

In loving memory of my father,

Mr. G. M. T. Emede.

Mathematics N5

Mathematics N3

Mathematics N2

In loving memory of my father,
Mr. G. M. T. Emede

Preface

The first chapter of this textbook has been written to explain in the detail graphs of various functions and their corresponding defined and undefined regions.

It is usually very difficult for FET College students to solve equations of functions when the learner in question does not have a fundamental knowledge of the range and domain of a specific function.

The principle of Complex Numbers has been applied to solve Engineering Science and Electrical Engineering problems in Chapter 5 of this book.

The concept of differentiation has also been defined from trigonometric, direct and indirect proportionality perspectives in Chapter 6. This is a very basic concept for students to understand because they will be applying this knowledge in Applications of Differentiation, Engineering Mathematics N5.

The last chapter of this book is Integration. A practical and logical approach has been applied in teaching the limit of integration or definite integrals.

Table of Content

Chapter 1 Functions and Graphs

1.1 Quadratic Functions The standard form of a quadratic function is $y = ax^2 + bx + c$ $c = y$ intercept of the curve $b/2a = $ line of symmetry of the curve The x intercepts or the roots of the curve can be calculated by: a) Factorizing b) Applying Quadratic Formula c) Completing the square **1.1.1 Domain and Range of a Quadratic Function.** Consider the two curves and tables of values as shown below.

Table 1.1 $y = x^2$

x	-1	-2	-3	-4	-5
y	1	4	9	16	25
x	+1	+2	+3	+4	+5
y	1	4	9	16	25

Table 1.2 $y^2 = x$

x	-1	-2	-3	-4
y	1	4	9	16
x	+1	+2	+3	+4
y	1	4	9	16

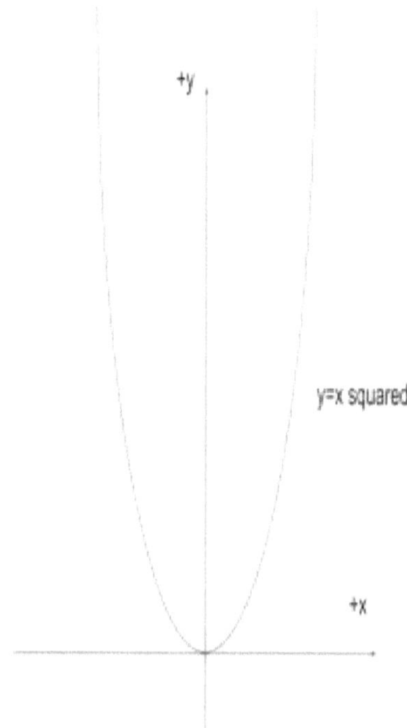

Fig 1.1 The graph of $y=x^2$

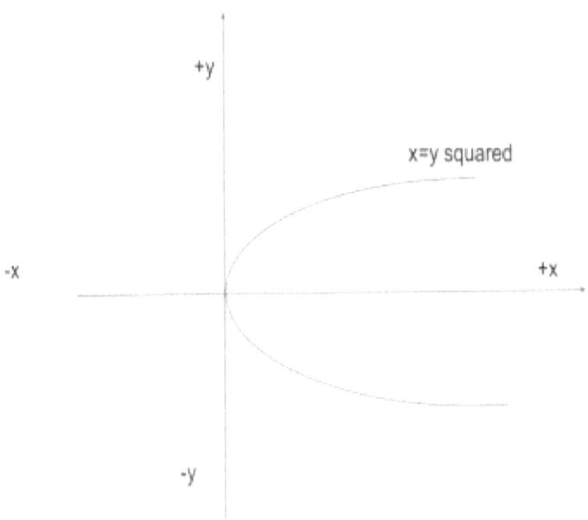

Fig 1.2 The graph of $x=y^2$

For the equation, $y^2 = x$, there are two values of y for each given value of x (+5 and -5 for x = 25 etc). This equation is a relation.

Considering the curve $y = x^2$, there are two values of x for each given value of y (+3 and -3 for y = 9). This curve is a function. Please note that a vertical line will intersect a relation at two points while a vertical line will intersect a function at one point.

The Domain of a curve specifies all the possible values of the x coordinates.

The Range of a curve specifies all the possible values of the y coordinates.

Example 1 State the Domain and Range of the two curves

i. $y^2 = x$ domain $= 0 \leq x \leq + \infty$ range $= - \infty \leq y \leq + \infty$

ii. $y = x^2$ domain $= - \infty \leq x \leq + \infty$ range $= 0 \leq y \leq + \infty$

The term $0 \leq x$ means that x is greater than or equal to 0. Likewise, $x \leq + \infty$ means that x is less than or equal to plus infinity.

1.1.2 Classification of Quadratic Roots Quadratic curves are also referred to as parabolas. The general formula for a parabola is $y = ax^2 + bx + c$.

a = coefficient of x^2 b = coefficient of b^2 and c = y intercept c is the value of y when x is zero.

The roots of the equation or the x intercepts of the equation can be evaluated by applying the methods as stated in section 1.1. In this section, we shall apply the quadratic formula as stated below.

$$\text{Quadratic formula} \quad x = \frac{-b \pm \sqrt{b^2 - 4ac}}{2a}$$

The line of symmetry of a parabola is given by the formula: $x = \frac{-b}{2a}$ where a = coefficient of x^2

And b = coefficient of x and c is the constant term.

Substituting this value of $x = \frac{-b}{2a}$ into the original equation will give the corresponding value of y. These values of x and y are the coordinates of the turning points of the parabola. The above methods of evaluating the roots of x intercepts of a quadratic curve is explained in examples 7 - 9 of Section 2.3. The Fig. 1.2 shows graphs of : a) $y = x^2 - 4x + 4$ Repeated roots b) $y = 7x^2 - 23x + 6$ Real unequal roots c) $y = x^2 - 5x + 6$ Real unequal roots d) $y = x^2 - 5x + 15$ Quadratic Complex Number

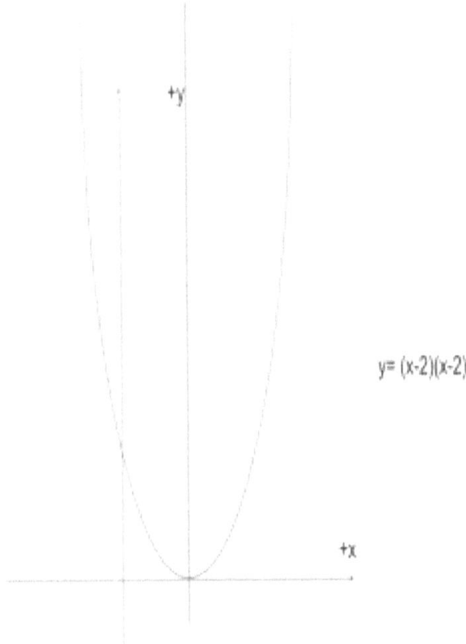

Fig 1.3 Classification of quadratic roots. $y = x^2 - 4x + 4$ Repeated roots

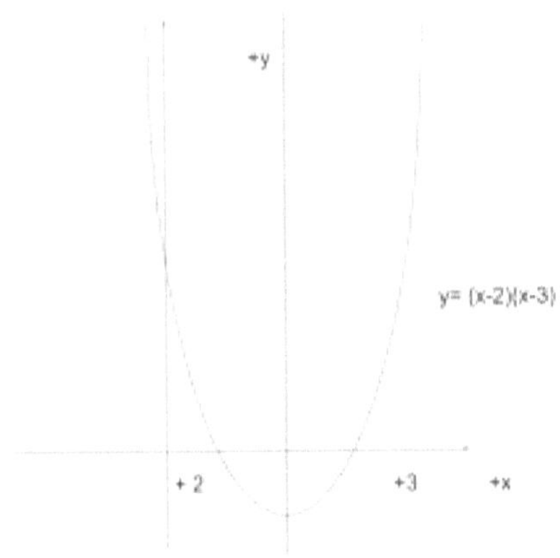

Fig   1.4

Classification of quadratic roots. $y = 7x^2 - 23x + 6$ Real unequal roots

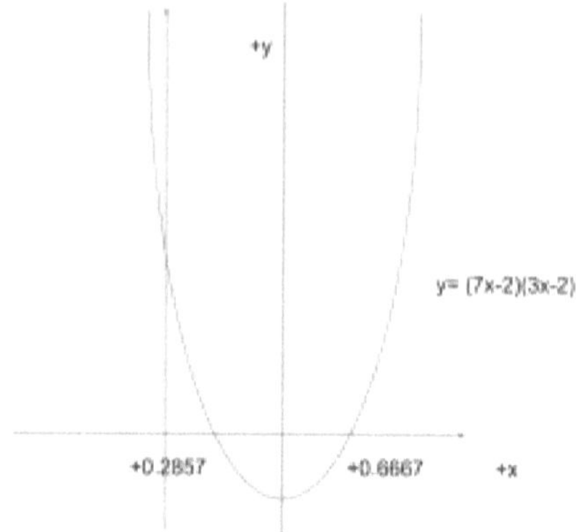

Fig 1.5 Classification of quadratic roots. $y = 7x^2 - 23x + 6$ Real unequal roots

Fig 1.6 Classification of quadratic roots. $y = x^2 - 5x + 15$. Quadratic Complex Numbers.

Table 1.3

	Factorized Form	Roots	$[b^2 - 4ac]^{0.5}$
$y = x^2-4x + 4$	$y = (x-2)(x-2)$	$x = -2$; $x = -2$	$[(-4)^2 - 4(1X4]^{0.5}$ 0
$y = x^2-5x + 6$	$y = (x-3)(x-2)$	$x = 3$; $x = 2$	$[(-5)^2 - 4(1X6]^{0.5}$ 1
$y = 7x^2-23x + 6$	$y = (7x-2)(3x-2)$	$x = 2/7$; $x = 2/3$	$[(-23)^2 - 4(7X6]^{0.5}$ 19
$y = x^2-5x + 15$	NA		$[(-5)^2 - 4(1X15]^{0.5}$ error

The term, $[b^2 - 4ac]^{0.5}$ for the equation, $x^2-5x + 15$, is the square root of a negative number. This is the reason why our calculators give an undefined or error value. Furthermore, this curve does not have x intercepts or roots. The equation, $y = x^2-5x + 15$, is a Quadratic Complex number as stated above.

Example 1: Given the equation $y = 6x^2 + 11x + 3$, draw the graph to scale by calculating the following:

    a. y intercept b) the roots c) the turning points

Solution:

    a. when $x = 0$, $y = 6(0)^2 + 2(0) + 3 = 3$ x, y $= (0,3)$

    b. Factorizing $(3x +1)(2x +3) = 0$ $x = -\frac{1}{3}$; $x = -\frac{3}{2}$ x, y $= (-\frac{1}{3}, 0$) ; $(-\frac{3}{2}, 0)$

    c. $X = -\frac{b}{2a} = -\frac{11}{2(6)} = -0.917$

$y = 6(-0.917)^2 + 11(-0.917) + 3 = -2.042$ x, y = (-0.917, -2.042)

(-0.917, -2.042)

Fig 1.7 Graph of $y = 6x^2 + 11x + 3$

1. Continuous and Discontinuous Functions

Consider the two functions $y = x^2$ and $y = \frac{8}{x}$. For the equation, $y = x^2$, substituting any possible x value that lies within the domain gives a corresponding y value. This is not correct for the equation, $y = \frac{8}{x}$. The domain for the function is $0 < x \leq + \infty$ and $0 < x \leq - \infty$. Compare this domain with the single expression for the domain for $y = x^2$. This means that the values of x =0 and y = 0 does not exist. $y = \dfrac{8}{0} =$ error or

undefined when x = 0 and x = $\dfrac{8}{0}$ = error or undefined when y = 0.

The lines x = 0 and y = 0 are known as asymptotes because they are the lines of discontinuity for the given function. This means that the lines x=0 and y=0 are undefined for the function y= $\dfrac{8}{x}$. Fig 1.4 and table 1.4 illustrates the asymptotes x=0 and y=0. An asymptote is a line that differentiates a defined region from an undefined region or area. Compare the curves y1 and y2 that are in the 1st and 3rd quadrant of the Cartesian plane and the curve y = x^2. (See Fig 1.4)

y= $\dfrac{8}{x}$ is a discontinuous function while y = x^2 is a continuous

function. y= $\dfrac{8}{x}$ is discontinuous at the lines x = 0 and y = 0.

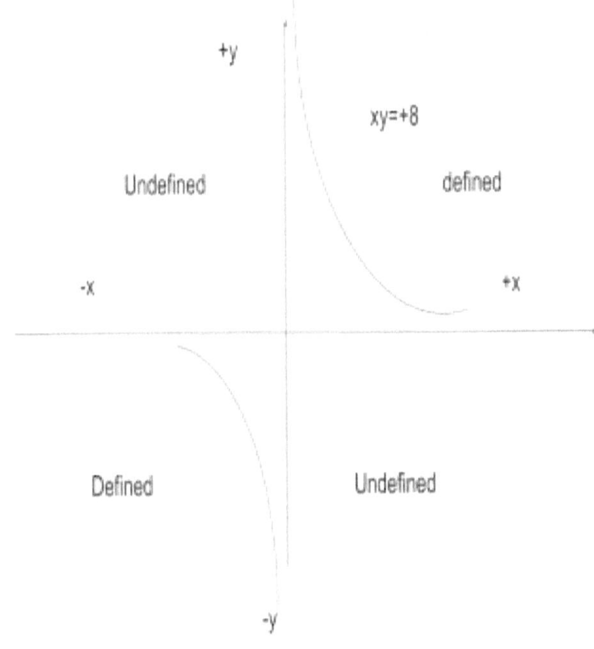

Fig 1.8 The Hyperbolic Curve or the hyperbolic function $xy=+8$
Table 1.4

x	-4	-3	-2	0	2	3	4
y	-2	-2.67	-4	undefined	4	2.67	2

1.3 Inverse Functions The inverse of a function is obtained by interchanging the values of y with x and vice versa. The inverse of $y = x^2$ is $x = y^2$. See Fig 1.5, Table 1.5 and Table 1.6.

Careful consideration should be given to the replacement of the y values with x values and vice versa.

1.3.1 Inverse of a Quadratic Function. For the sake of simplicity, we will be plotting the graph of $y=x^2$ and its inverse. The same procedures will be taken as in the section 1.1.1. The inverse of $y=x^2$ is $x=y^2$.

The mirror line test of $y=x$ has been applied. This test confirms that our inverse function is correct.

Table 1.5 Table of values of $y=x^2$

x	-6	-5	-4	-3	-2	-1	0
y	36	25	16	9	4	1	0
x	1	2	3	4	5	6	7
y	1	4	9	16	25	36	49

Table 1.6 $x=y^2$

x	36	25	16	9	4	1	0
y	-6	-5	-4	-3	-2	-1	0
x	1	4	9	16	25	36	49
y	1	2	3	4	5	6	7

1. **Lines of Symmetry** A line of symmetry is a line that bisects a curve into 2 identical halves. If the curve is folded along this line of symmetry, the two halves

will overlap because they have identical shape. A line of symmetry can also be called the mirror line of a curve. This is because if you place a mirror along this line and look into the mirror, you will see the other half that completes curve. The line of symmetry of y $= x^2$ is the line x $= 0$ or the y axes. The line of symmetry of $y^2 = x$ is the line y $= 0$ or the x axes. See Fig 1.9 and Fig. 1.4.

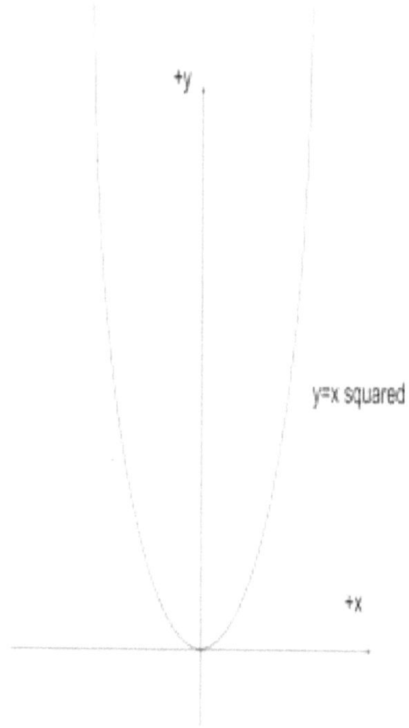

Fig 1.9 The line of symmetry defined by $y = -b/(2a.$ $y = x^2$

1.5 The Straight Line A straight line is the intersection of two planes. The vertical or horizontal sharp edge of a door or a table is a straight line. The following standard equations are used for straight-line graphs.

1. Gradient form $y = mx + c$ $m = \tan\Theta$ and m = gradient or slope c = y intercept
2. General form $ax + by + c = 0$ c = y intercept
3. Intercept form $\frac{x}{a} + \frac{y}{b} = 1$ a = x intercept and b = y intercept.
4. Point Gradient form $y - y_1 = m(x - x_1)$ m = slope
5. Two Point form $(y_2 - y_1) \div (x_2 - x_1) = m$

1.5.1 Parallel Lines

Two parallel lines $y_1 = m_1 x_1 + c_1$ and $y_2 = m_2 x_2 + c_1$ have the same slope $m_1 = m_2$. However, they do not have the same y intercept. See Fig 1.6.

1.5.2 Perpendicular Lines Two lines $y_1 = m_1 x_1 + c_1$ and $y_2 = m_2 x_2 + c_1$ are perpendicular if the product of their slopes is -1. $m_1 m_2 = -1$.

1.5.3 Angle of Inclination of a Line Since $m = \tan\Theta$, the angle of inclination $\Theta = \tan^{-1}(m)$

1.5.4 Distance between Two Points Calculation of the distance between two points, A and B, is carried out by applying Pythagoras Theorem. This is because all inclined lines make a right-angled triangle when a vertical line is drawn from the higher point on the line and a horizontal line is drawn from the lower point of the inclined line. For the following examples assume the following notation: $A = x_A, y_A$, $B = x_B, y_B$ etc The distance between two points is given as: $[(x_B - x_A)^2 + (y_B - y_A)^2]^{0.5}$ Examples 1-8 below are based on Fig 1.6 below as shown.

Example 1 Find the equation of line AB in the standard form $y = mx + c$ given that A (0, 10) and B (2, 8).

Slope $m = \dfrac{8-10}{2-0} = -1$ and $c = 10$ $y = -x + 10$

Example 2 Find the equation of line EDB in the intercept form $\dfrac{x}{a} + \dfrac{y}{b} = 1$ given that E = (-5, 0) and D= (0,8). intercept a = -5 and y intercept, b = 8. Equation of the line is $\dfrac{x}{-5} + \dfrac{y}{8} = 1$

Example 3 Prove that the triangle ADB is a right angle triangle if A= (0, 10) , B= (2,8) and D = (0,6). Slope of line AB = -1 (Example 1)

Slope of line DB = $\dfrac{8-6}{2-0} = 1$

For perpendicular lines $m_1 m_2 = -1$ therefore $1 \times -1 = -1$ The angle ADB is a right angle triangle. Example 4 Prove that the lines DF and

AG are parallel given that D= (0,6), F = (6,0) , A (0, 10) and G = (10,0).

Slope of line DF = $\dfrac{6\text{-}0}{0\text{-}6}$ = -1 Slope of line AG = $\dfrac{10\text{-}0}{0\text{-}10}$ = -1 Lines DF and AG are parallel because they have the same slope or gradient. Example 5 Evaluate the angle of inclination of the following lines:

   i.   AB, A (0, 10) , B (2,8)

  ii.   DB, B (2,8) D (0,6)

 iii.   IJ given that I = (2.5, 7.5) and J = (-0.5, 0)

   i.   Slope m_{AB} = -1 = tan Θ Θ = tan^{-1} -1 = 135^0

  ii.   Slope m_{DB} = 1 = tan Θ Θ = tan^{-1} 1 = 45^0

 iii.   Slope m_{IJ} = $\dfrac{7.5\text{ -}0}{2.5\text{ -}\,-0.5}$ = 2.5 = tan Θ Θ = tan^{-1} 2.5 = 68.19^0

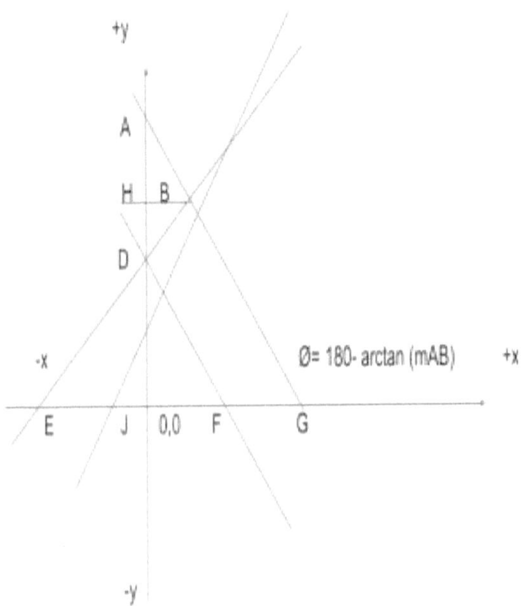

Fig 1.10 Examples 1-8 straight-line problems.

Example 5 Evaluate the angle of inclination of the following lines:

   i.   AB, A (0, 10) , B (2,8)

  ii.   DB, B (2,8) D (0,6)

 iii.   IJ given that I = (2.5, 7.5) and J = (-0.5, 0)

   i.   Slope m_{AB} = -1 = tan Θ Θ = tan^{-1} -1 = 135^0

  ii.   Slope m_{DB} = 1 = tan Θ Θ = tan^{-1} 1 = 45^0

 iii.   Slope m_{IJ} = $\dfrac{7.5 - 0}{2.5 - -0.5}$ = 2.5 = tan Θ Θ = tan^{-1} 2.5 =

68.19^0

Example 6 Calculate the distance between the two points :

   i. A ((0, 10) and B (2,8).
  ii. D (0,6) and B(2,8)
 iii. I (2.5, 7.5) and J (-0.5, 0)

   i. $[(x_B - x_A)^2 + (y_B - y_A)^2]^{0.5} = [(2 - 0)^2 + (8 - 10)^2]^{0.5} =$ 2.828

or $[(x_A - x_B)^2 + (y_A - y_B)^2]^{0.5} = [(0 - 2)^2 + (10 - 8)^2]^{0.5} =$ 2.828

   i. $[(x_D - x_B)^2 + (y_D - y_B)^2]^{0.5} = [(0 - 2)^2 + (6 - 8)^2]^{0.5} = 2.828$

$[(x_B - x_D)^2 + (y_B - y_D)^2]^{0.5} = [(2 - 0)^2 + (8 - 6)^2]^{0.5} =$ 2.828

   i. $[(x_J - x_I)^2 + (y_J - y_I)^2]^{0.5} = [(2.5 - - 0.5)^2 + (7.5 - 0)^2]^{0.5} =$ 8.078

$[(x_I - x_J)^2 + (y_I - y_J)^2]^{0.5} = [(-0.5 - 2.5)^2 + (0 - 7.5)^2]^{0.5} = 8.078$

The equation $[(x_B - x_A)^2 + (y_B - y_A)^2]^{0.5}$ is a modification of the Pythagoras Theorem. A right angle triangle is made up of two components namely the base and the height. $x_B - x_A$ = base of the right angle triangle $y_B - y_A$ = height of the right angle triangle

Example 7. Evaluate the slope of the line $\dfrac{x}{-5} + \dfrac{y}{8} = 1$.

Write the equation in the form $y = mx + c$ $\frac{y}{8} = 1 + \frac{x}{5}$ (multiply both sides by 8)

$Y = 1.6x + 8$ Thus, m = 1.6

Example 8 Write the equation of the line IJ, I (2.5, 7.5) and J (-0.5, 0),

the form $y = mx + c$ by applying the two point formula.

$(y_J - y_I) \div (x_J - x_I) = (y_I - y_J) \div (x_I - x_J) = (0\ \text{-}7.5) \div (\text{-}0.5\ \text{-}2.5) = 2.5 = m$

a. $= (y - y_J) \div (x - x_J) = (y - y_I) \div (x - x_I)$ Substituting values of point J ; $2.5 = (y\ \text{-}0) \div (x - \text{-}0.5)$

$y = 2.5x + 1.25$

1.5.5 Inverse of a Straight Line Given the function $y = 3x + 10$, we can find the inverse by replacing y with x and vice versa: $x = 3y + 10$ making y the subject of the formula: $y = \frac{1}{3}(x - 10)$ The next step will be completing a table of values for both lines.

Table 1.7 $y = 3x + 10$ $y = \frac{1}{3}(x - 10)$

X	0	-3.333	x	0	10
y	10	0	y	-3.333	0

The line y=x is a mirror line. This means that by placing a mirror on the line y=x facing the line $y = \frac{1}{3}(x - 10)$, the line y= 3x + 10 is seen in the mirror.

Likewise, placing the mirror on the mirror line such that it faces the line y= 3x + 10 shows us an image of $y = \frac{1}{3}(x - 10)$ in the mirror.

0,10 y y = 3x +10 y = x

-3.333,0 0,0 y = $\dfrac{1}{3}$ (x -10)

10,0

0, -3.333

Fig 1.11 The graph of y= 3x + 10 and its inverse.

Exercise 1.1

1. Draw a graph of the following curves by completing a table of values for the following: a) y= 10/x b) y = 2/x c) x= -3/y d) x = -5/y

2. State the equation of the line in the form y = mx + c given the following:

a. Passes through point (0,5) with a slope of 2.
b. Passes through the point (5,8) with a gradient of -1.
c. Intersects the x axis at -1 with a slope of 0.5.
d. Intersects the y axis at the point -5 with a slope of 1.

e. Passes through the point (4, -2) and $\Theta = 20^0$.

f. Passes through the point (3, 1) and $\Theta = 80^0$

1. Draw the lines stated in Question 1 a) – f) on the same axes or otherwise by using a suitable scale.

Label the following for each line:

a. Point through which it passes in the form (x,y).
b. Angle of inclination
c. Draw and label the horizontal and vertical distances between any two points on each line.

1. Find the equation of the lines passing through the points by applying the formula:

$$(y_2 - y_1) \div (x_2 - x_1) = (y_3 - y_2) \div (x_3 - x_2) = m.$$ Your equation should be in the form

$$y = mx + c.$$

a. $(2,-2)$ and $(3, 4)$
b. $(10,10)$ and $(-2, -7)$
c. $(7,6)$ and $(-3, -3)$
d. $(1,10)$ and $(2,5)$
e. $(-2,-3)$ and $(-8,-9)$
f. $(-20,-9)$ and $(-6, 7)$

1. Find the angle of inclination and gradient of the lines stated below.

a. $\dfrac{x}{12} + \dfrac{y}{2} = 5$
b. $\dfrac{6x}{7} + \dfrac{y}{8} = 3$
c. $10x - 2y + 16 = 0$
d. $2x - 15y - 20 = 2$

e. $y-6 = 10x-2$

f. $y-15=6x-10$

1. Find the equation of the lines perpendicular to the lines stated below.

a. $\frac{x}{1} + \frac{y}{2} = 3$ passing through point 1,1

b. $\frac{x}{3} - \frac{y}{4} = 3$ passing through point -2, -1

c. $2y = 6x-12$ passing through -2, 1

d. $6x - 3y - 18 = 0$ passing through 2, -1

e. $y-6 = 4x-2$ passing through the origin

f. $y-18 = -3x-10$ passing through 5,5

1. Find the equations of the six lines parallel to the lines stated in Question 5 a) – f) passing through the point (6,7).
2. Find the equations of the six lines perpendicular to the lines stated in Question 5 a) – f) passing through the point (2,-1).
3. Find the equations of the six lines parallel to the lines stated in Question 5 a) – f) passing through the point (-2,-3).
4. Find the equations of the six lines perpendicular to the lines stated in Question 5 a) – f) passing through the point (-1,2).

1.6 The Circle The General Formula for the circle is $x^2 + y^2 = r^2$, where r is the radius of the circle, x is the x coordinate of any point on the circle and y is the corresponding y coordinate of the same point. This equation of a circle is based on the fact that the center of the circle is at the origin (0,0) and the radius, r, is a constant distance measured from the center to any point on the circle, x,y.

The classification of the circle is based on the semi circles as illustrated in the diagrams below.

The first and second diagrams are based on the positive and negative semicircles. The positive and negative symbol is placed or written on the RHS of the equation when x is made the subject of the formula. Another very important concept is that the expression or the formula on the LHS is in terms of y.

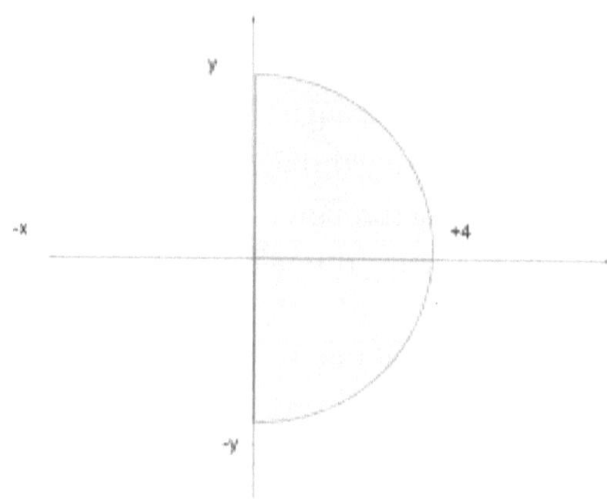

Fig 1.12 The Semicircle on the positive x axis. $x=+(r^2-y^2)^{0.5}$

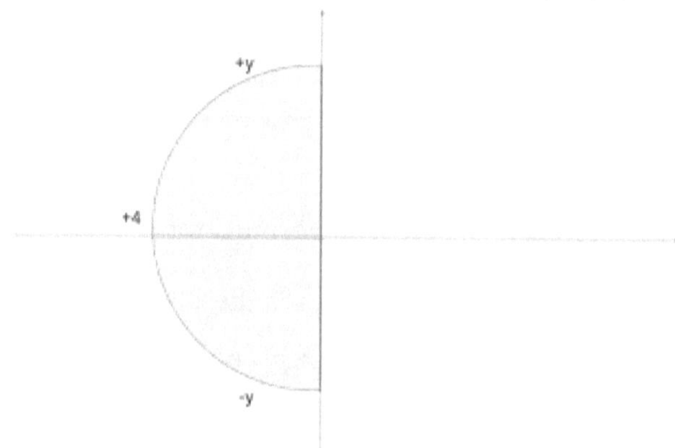

Fig 1.13 The Semicircle on the negative x axis. $x=-(r^2-y^2)^{0.5}$

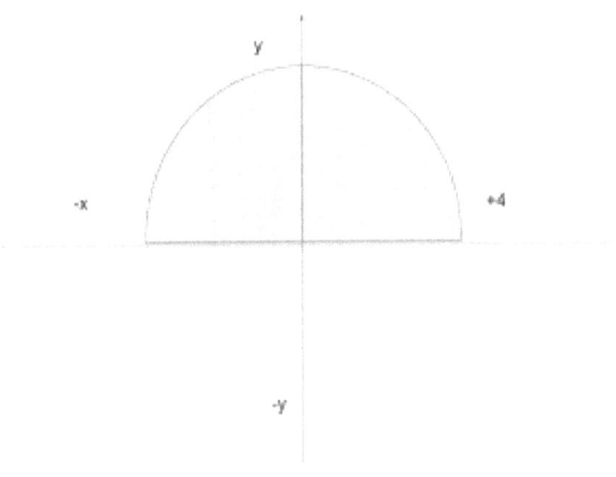

Fig

1.14 The Semicircle on the positive y axis. $y=+(r^2-x^2)^{0.5}$

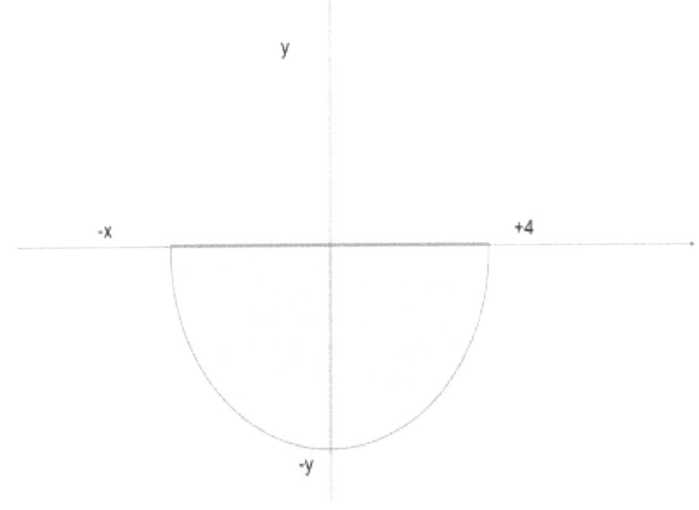

Fig 1.15 The Semicircle on the negative y axis. $y=-(r^2-x^2)^{0.5}$

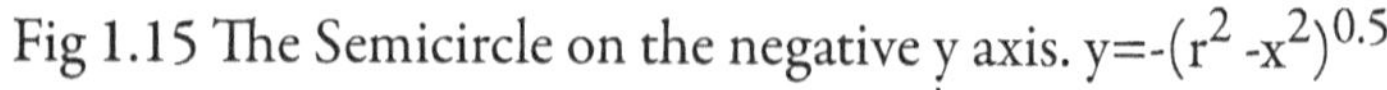

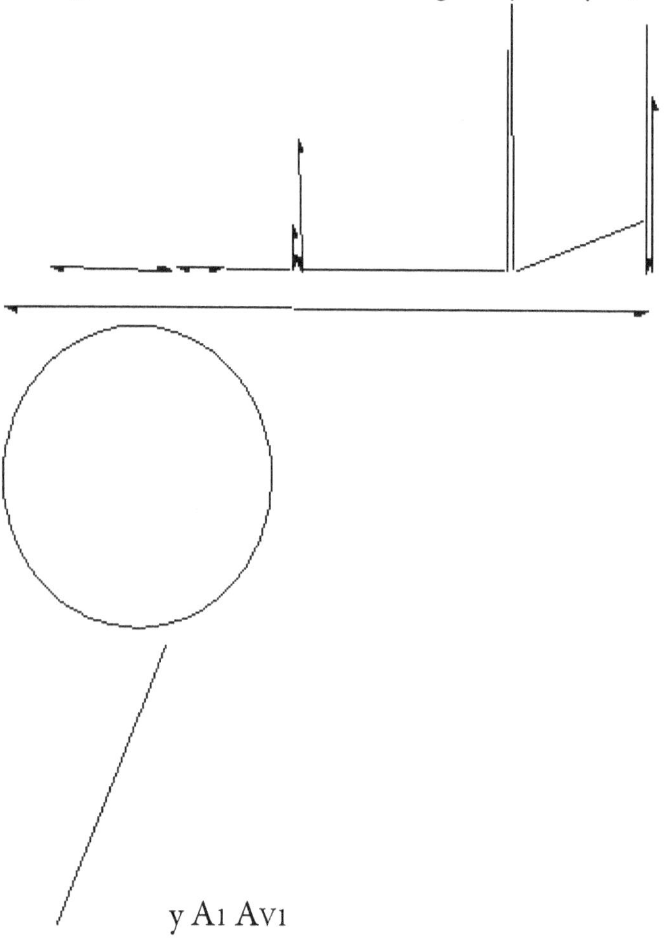

Fig 1.16 The Circle

Consider the two points A and B making an angle of 60 and 30 degrees with the positive x axes respectively. Assuming that the radius r = 1. Applying basic trigonometry,

A_V = Vertical height of A above the x axes = $(1) \sin 60 = 0.866$

B_V = Vertical height of B above the x axes = $(1) \sin 30 = 0.5$

A_H = Horizontal distance of A from the origin = $(1) \cos 60 = 0.5$

B_H = Horizontal distance of B from the origin = $(1) \cos 30 = 0.866$.

Substituting into the formula for a circle:

Point A: $(0.866)^2 + (0.5)^2 = 0.9999 = 1^2$

Point B: $(0.5)^2 + (0.866)^2 = 0.9999 = 1^2$

Example 1 Prove that the equation of the unit circle (Fig 1.6) above the x axis is $y = +(r^2 - x^2)^{0.5}$.

Consider the point A_1 with coordinates as stated in the diagram, r = 1, $A_{H1} = 0.5$ and

$A_{V1} = 0.866$. Substituting these values into the equation, we have

$$y = -(1^2 - (0.5)^2)^{0.5} = +0.866 = A_{V1}$$

Example 2 Evaluate the radius of the equation $y = +(81 - x^2)^{0.5}$.

Write the equation in the standard form $x^2 + y^2 = r^2$

Square both sides: $y^2 = 9^2 - x^2$ $y^2 + x^2 = 9^2$ radius, r = 9

Example 3 Given the equation of a circle, $0.0625\,x^2 + 0.0625y^2 = 1$. Calculate:

    a. The domain and range of the graph.
    b. Is it a continuous or discontinuous function?
    c. Write the equations of four lines of symmetry.
    d. State whether it is a function or a relation. Solution

    a. $X^2 + y^2 = 1 \div 0.0625$ $x^2 + y^2 = 16$ $x^2 + y^2 = 4^2$ domain = $-4 \le x \le +4$; range = $-4 \le y \le +4$.

b. Continuous function

c. $X = 0$; $y = 0$ $x = y$; $x = -y$.

d. A relation

Exercise 1.2

1. Evaluate radius, r, and state the domain and range of each of the following:

a. $x^2 + y^2 = 0.5$

b. $x^2 + y^2 = 10000$

c. $x^2 + y^2 = 3600$

d. $x^2 + y^2 - \dfrac{3}{4} = 0$

e. $x^2 + y^2 - \dfrac{9}{4} = 0$

f. $x^2 = \dfrac{6}{7} - y^2$

1. Sketch and label the circles as stated below.

a. $x^2 + y^2 = 0.5$

b. $x^2 + y^2 = 10000$

c. $x^2 + y^2 = 3600$

d. $x^2 + y^2 - \dfrac{3}{4} = 0$

e. $x^2 + y^2 - \dfrac{9}{4} = 0$

f. $x^2 = \dfrac{6}{7} - y^2$

1.7. Ellipse

General Equation for the ellipse is: $\dfrac{x^2}{a^2} + \dfrac{y^2}{b^2} = 1$

a = x intercept b = y intercept

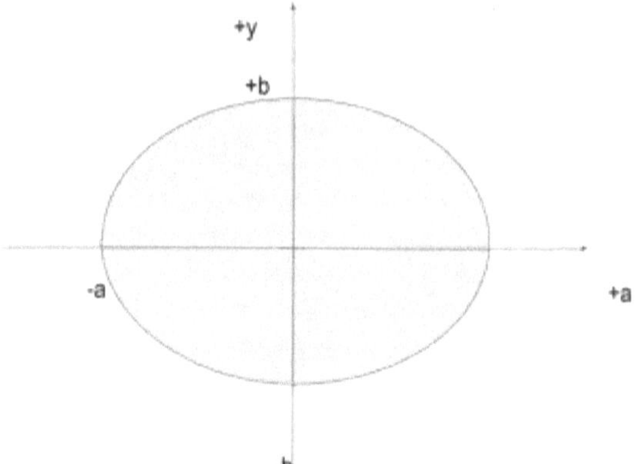

+y
+b
-a
+a
-b

Fig. 1.17 the Ellipse

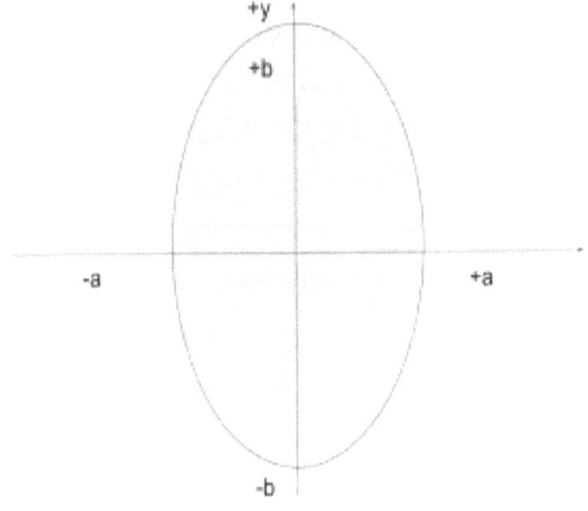

Fig. 1.18 The Ellipse

Example 1 Given the equation $\dfrac{x^2}{16} + \dfrac{y^2}{9} = 144$

a. Write the equation in the standard form $\dfrac{x^2}{a^2} + \dfrac{y^2}{b^2} = 1$.
b. State the value of a and b.
c. State the domain and the range.
d. State the equations of lines of symmetry.
e. State whether it is continuous or discontinuous.
f. Sketch and label the ellipse.

Solution:

a. $\dfrac{1}{144} \left[\dfrac{x^2}{16} + \dfrac{y^2}{9} \right] = \dfrac{144}{144}$ $\dfrac{x^2}{2304} + \dfrac{y^2}{1296} = 1$ and $\dfrac{x^2}{48^2} + \dfrac{y^2}{36^2} = 1$
b. a= 48 ; b = 36
c. Domain = -48 ≤ x ≤ + 48 Range = -36 ≤ y ≤ + 36.
d. X = 0 and y = 0.
e. Continuous

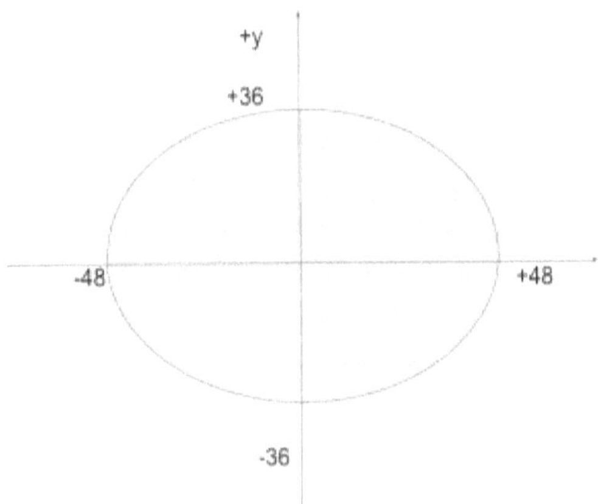

Fig 1.19 Ellipse, $\dfrac{x^2}{48^2} + \dfrac{y^2}{36^2} = 1$

Exercise 1.3

Sketch and label the x and y intercepts of the following ellipses.

1. $\dfrac{x^2}{7} + \dfrac{y^2}{15} = 1$

2. $\dfrac{x^2}{17} + \dfrac{y^2}{24} = 1$

3. $\dfrac{x^2}{a^2} + \dfrac{y^2}{b^2} = 1$

4. $64x^2 + 9y^2 = 576$

5. $25x^2 + 64y^2 = 1600$

6. $25x^2 + 9y^2 - 225 = 0$

7. $9y = [1 - 64x^2]^{0.5}$

8. $y = \left[\dfrac{1 - 100x^2}{64}\right]^{0.5}$

9. $64x^2 + 9y^2 = 1$

10. $25x^2 + 64y^2 = 1$

1. Hyperbolas

1.8.1 Rectangular Hyperbola

The figure below illustrates the development of conic sections of hyperbolas, circles, ellipse and parabolas.

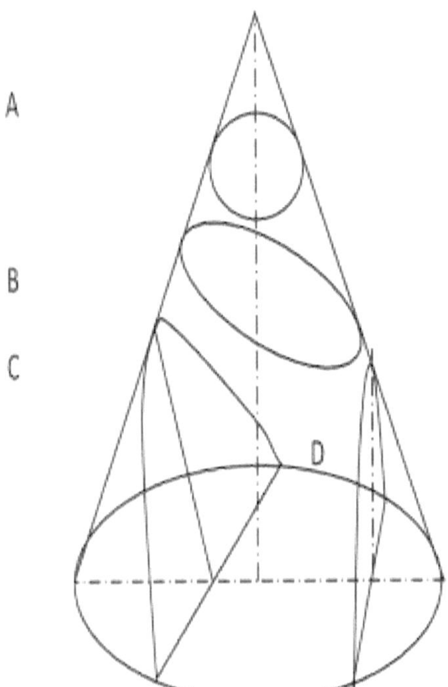

Fig. 1.20 Development of Conic Sections

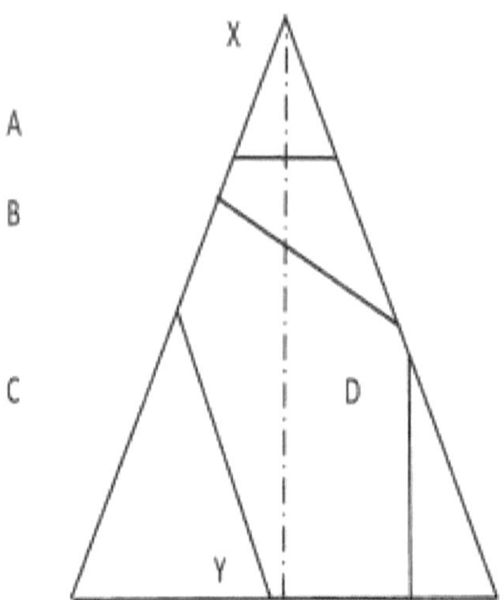

Fig. 1.21 Development of Conic sections

These curves can be classified based on the angle that the cutting plane makes with the vertical line XY.

Table 1.8

Conic Section	Cutting Plane Angle	Intersecting line XY
A Circle	90	Yes
B Ellipse	Acute	Yes
C Parabola		No
D Hyperbola	Parallel	No

The hyperbola formed by the cutting plane D above forms a curve given by the equation $\frac{x^2}{a^2} - \frac{y^2}{b^2} = 1$. The rectangular hyperbola $y = k/x$ is a curve developed when the Cartesian plane $x - y$ is rotated -45^0 such that the asymptotes of the curve and the x and y axes are one and the same line respectively. Secondly, the value of a is equivalent to b.

Example 1 Draw the rectangular hyperbola $xy = 8$ by completing a table of values and choosing a suitable scale for the x and y axes. Table 1.9

x	-4	-3	-2	0	2
y	-2	-2.67	-4	error	4
K	8	8	8		8

x axes 1 unit = 10mm, y axes 1 unit = 20mm

Example 2 Complete a table of value for $xy = -1$. Draw the curve by choosing a suitable scale. Table 1.10

x	-4	-3	-2	0
y	0.25	0.33	0.5	error
$k = xy$	-1	-1	-1	

x axes 1 unit = 10mm, y axes 1 unit = 20mm

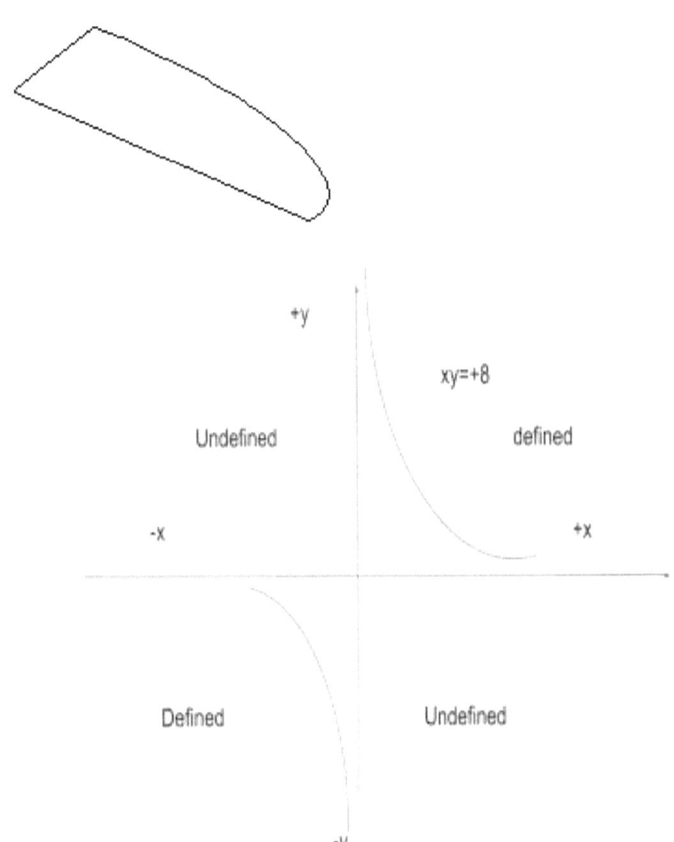

+y
xy=+8
Undefined
defined
-x
+x
Defined
Undefined
-y

Fig. 1.22 The Rectangular Hyperbola

1.8.2 Non Rectangular Hyperbola.

$\dfrac{x^2}{a^2} - \dfrac{y^2}{b^2} = 1$ where a = x intercept and b = y intercept.

The asymptotes of the curve are given by

a. $y = \dfrac{b}{a} x$ compare with the straight line equation $y = mx + c$
    where c = 0.

b. $y = -\dfrac{b}{a} x$ and $y = +\dfrac{b}{a} x$ are slopes of the asymptotes.

Example 1 Evaluate the following and draw the graph of

$\dfrac{x^2}{9^2} + \dfrac{y^2}{6^2} = 1$

a. x intercepts b) The asymptotes c) The lines of symmetry

Solution:

a. (3, 0) and (-3, 0) b) $Y = \dfrac{6}{9} x$ and $y = -\dfrac{6}{9} x$ c) x axes and y axes.
    See Fig. 1.14

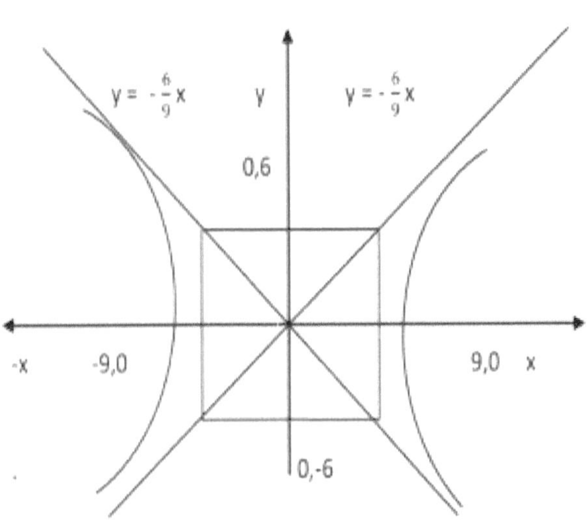

Fig 1.23 Non Rectangular Hyperbola

Exercise 1.4

1. Sketch the following rectangular hyperbolas by completing a table of values and choosing a suitable scale for the x and y axes.

a. $xy = -9$

b. $y = \dfrac{30}{x}$

c. $xy = -7$

d. $x = \dfrac{36}{y}$

1. Sketch and label the following non-rectangular hyperbolas by evaluating the equation of the asymptotes.

a. $\dfrac{x^2}{121} - \dfrac{y^2}{64} = 1$

b. $25x^2 - 121y^2 = 3025$

c. $\dfrac{x^2}{81} - \dfrac{y^2}{36} = 1$

d) $81x^2 - 121y^2 = 9801$

1.9 Exponential Functions

The standard form for an exponential function is $y = a^x$. The value of a may be greater than 1: $y = 3^x$. The value of a may be less than 1: $y = [\frac{1}{3}]^x = 3^{-x}$

1.9.1 The Domain and Range of an Exponential Function The x axis or the line y=0 is an asymptote of the exponential function. This is evident when we substitute the value of y=0 into the equation $y=a^x$.

Furthermore, taking the logarithm of both sides of the equation $0=a^x$ will definitely give us an undefined or error value. In conclusion, the domain and range of an exponential equation is as follows: Domain: $-\infty \le x \le +\infty$ Range: y is not equal to zero : $y \ne 0$ y is greater than zero : $0 < y < +\infty$

Example 1 Draw to scale the graph of $y = 3^x$ and $y = [\frac{1}{3}]^x$ by completing a table of values. a) State the domain and range of each graph b) What line defines the asymptote of each graph? c) Are the graphs continuous or discontinuous? d) State reasons why the exponential graph is a function.

Table 1.11 $y = 3^x$

x	-3	-2	-1	0	1	2	3
$Y = 3^x$	0.037	0.111	0.333	1	3	9	27

Table 1.12 $y = [\frac{1}{3}]_x$

x	-3	-2	-1	0	1	2	3
$Y = [\frac{1}{3}]_x$	27	9	3	1	0.333	0.111	0.037

Solution:

a. $y = 3^x$: domain $-\infty \le x \le +\infty$ range $0 < y \le +\infty$ $y \neq 0$

$y = [\frac{1}{3}]_x$: domain $-\infty \le x \le +\infty$; range $0 < y \le +\infty$ $y \neq 0$

a. $y = 0$ or the x axes
b. continuous
c. There are no two y values for each specific x value.

1. **The inverse of an Exponential Function.** The inverse of $y = 3^x$ is $x = 3^y$. Completing a table of values for $x = 3^y$ is as shown in the table below. Refer to the table 1.9.2 for the table of values of $y = 3^x$.

Table 1.13

$x = 3^y$	3^{-3}	3^{-2}	3^{-1}	3^0	3^1	3^2	3^3
y	-3	-2	-1	0	1	2	3

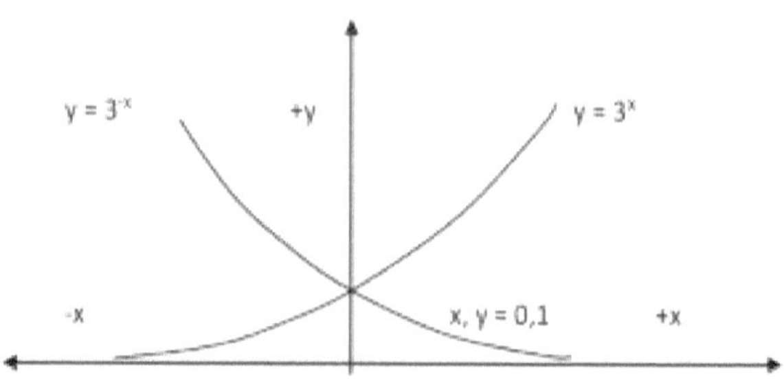

Fig 1.24 Graph of Exponential Functions $y = 3^x$ an $y = 3^{-x}$.

1.10 Logarithmic Functions

The general equation for a logarithmic function is $y = \log_a x$. The exponential form of this equation is written as $x = a^y$. Where a = base of the logarithmic function. Compare the $y = \log_a x$ with the following

1. $y = \log_3 81$
2. $y = \log_{10} 10000$
3. $y = \log_8 64$

Example 1 Draw the graph of $y = \log_4 4x$ and $y = \log_{1/4} 4x$ on the same axes

a. State the domain and range for each graph.
b. What line defines the asymptote of the graph?

Are both graphs continuous or discontinuous?

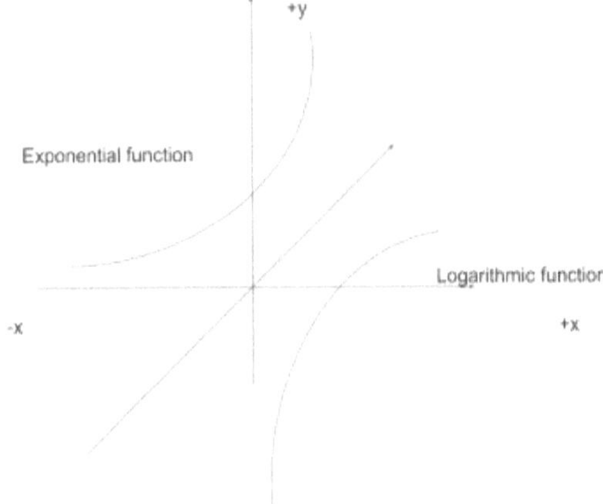

Fig 1.25 Exponential Function and its inverse

a. Table 1.14

x	4^{-3}	4^{-2}	4^{-1}	4^{0}
y	-2	-1	0	1

Table 1.15

x	4^{-3}	4^{-2}	4^{-1}	4^{0}
y	2	1	0	-1

Converting to exponential function: y= log 1/4 4x $[\frac{1}{4}]_{y}$

=4x 4x = $[\frac{1}{4}]_{-1}$ .x

a. y= log₄ 4x domain $0 \le x \le +\infty$ range $-\infty \le y \le +\infty$

y= log 1/4 4x domain $0 < x \le +\infty$ x ≠ 0 (x is not equal to 0) range $-\infty \le y \le +\infty$ b) y axes or x = 0.

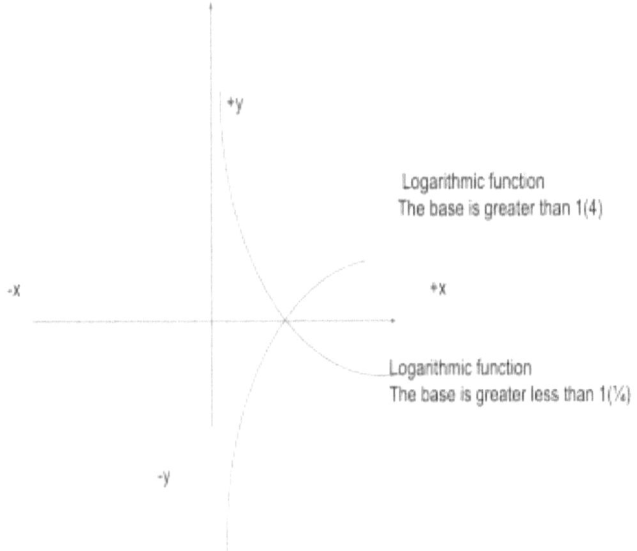

Fig. 1.26 Graph of Logarithmic Functions

   a. Continuous.

Example 2 Draw the graph of y = 10log1/4 4x by:

   a. Completing a table of values
   b. Choosing a suitable scale for both x and y axes
   c. State the x intercept of the graph from your table or graph.

The values of y can be evaluated by multiplying each y value of the example above by 10. Table 1.16

x	4^{-3}	4^{-2}	4^{-1}	4^0	4^1	4^2	4^3
y=10 log 1/4 4x	20	10	0	-10	-20	-30	-40

a. X axes : 10 units = 20mm

Y axes : 10 units = 10mm

a. X, y = $(1/4, 0)$

1.10.1 The domain and range of Logarithmic Function We are going to write the domain and range based on the tables and graphs of the two previous examples. Firstly, the y axis or the line x=o does not exist on the logarithmic graph. The y axis is an asymptote. Domain: $x \neq 0$ and $x \leq +\infty$ Range: $+\infty \leq y \leq +\infty$

1.10.2 The inverse of a Logarithmic Function See Section 10.1 The inverse of our basic logarithmic formula of $y = \log_a x$ is $x = \log_a y$. The table below explains the conversions from logarithmic functions to exponential functions and vice versa.

The equations $y = 3^x$ and $x = 3^y$ of Section 1.10 will be the used as examples.

Table 1.16

Function Inverse

Logarithmic Form $x = \log_3 y$　$y = \log_3 x$

Exponential Form $y = 3^x$　　$x = 3^y$

Comparing Fig 1.10.1 and Fig 1.10.2 will definitely confirm our conversions as stated on the table 1.10.2.

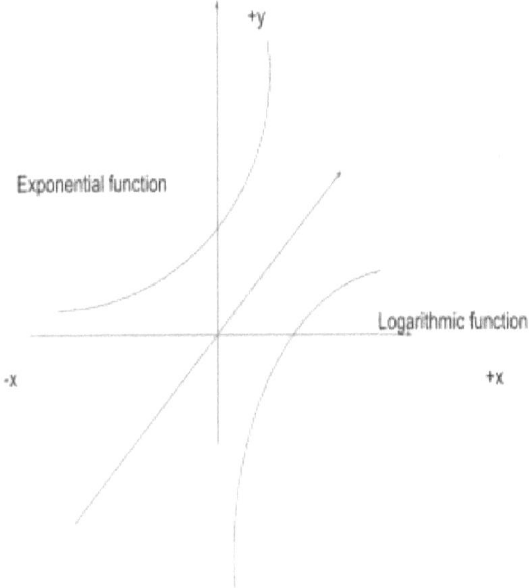

Fig 1.27 Logarithmic Graph and its inverse.

Exercise 1.5 1. Sketch and label the following curves by completing a table of values or otherwise. a) $y=3^x$

b) $y= 2(10^x)$

c) $y= (1/2)^x$

d) $y= 2\log x$

e) $y= 15\log_{1/2} x$

f) $y= 5\log 2x$

g) $y= \log_2 x$

h) $y= (0.1)^x$

1.11 Trigonometric Functions

A very practical example of a sine curve is alternating current plotted against time. The simplest trigonometric graphs as stated below have been dealt with in our series Engineering Mathematics N3.

a) y= sin x b) y= cos x c) y= tan x In this section we shall focus on the graphs of y = cosec x, y = sec x, y = cot x and the graphical solutions of simultaneous trigonometric equations as stated in the following examples. The x and y coordinates of the intersection of the two curves is the solution of the simultaneous trigonometric equation.

Example1 Find the graphical solution of the trigonometric equations $0 \leq x \leq 360$:

Y = 3sin(2x +30) and y = 3sin 2x

a. Read off the x and y values of the points of
b. intersection of both curves.
c. State the period of both graphs.
d. State the amplitude of both graphs.
e. State the frequency of both graphs.
f. State the phase shift, leading and lagging curve.
g. Domain of both graphs.

Table 1.17 y = 3sin(2x +30)

x 0   30 60 90 120 150
y 1.5 3   1.5 -1.5 -3 -1.5
x 180 210 240 270 300 330
y 1.5 3   1.5 -1.5 -3 -1.5

Table 1.18 y = 3sin 2x

x	0	30	45	60	90	120
y	0	2.59	3	2.59	0	-2.59

x	135	150	180	210	225	240
y	-3	-2.59	0	2.59	3	2.59

x	270	300	315	330	345	360
y	0	-2.59	-3	-2.59	-1.5	0

$Y = 3\sin(2x + 30)$

$60 \leq x \leq 90$; $15^0 + 60^0 = 75^0$ $y = 2\sin(2(75) + 30) = 0$ x, y = (75, 0) $150 \leq x \leq 180$; $15^0 + 150^0 = 165^0$ $y = 2\sin(2(165) + 30) = 0$ x,y = (165, 0) $240 \leq x \leq 270$; $15^0 + 240 = 255^0$ $y = 2\sin(2(255) + 30) = 0$ x, y = (255, 0) $330 \leq x \leq 360$; $15^0 + 330 = 345^0$ $y = 2\sin(2(345) + 30) = 0$; x, y = (345,0) The above calculations are shown to explain the intermediate values of x as stated. Similarly, the intermediate values of x for the second table have been evaluated.

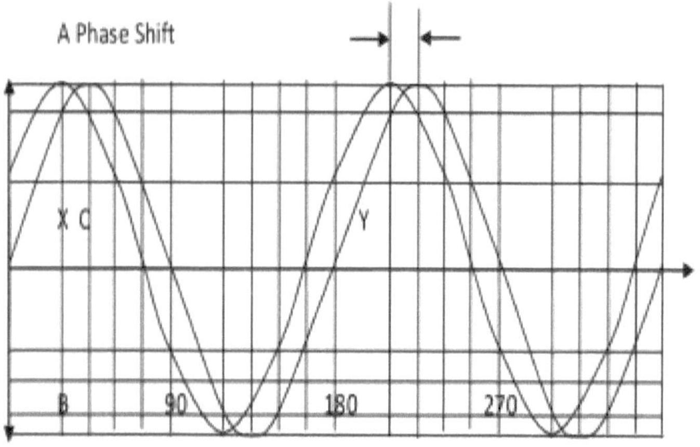

Fig. 1.28 Graph of y = 3sin(2x +30) and y = 3sin 2x

a) The points A, B and C are the y coordinates +3, -3 and 0respectively.x,y = 37.5, 2.898 ; 127.5, -2.898; 217.5, 2.898; 307.5, -2.898

.Note: Solving analytically $\tan 2x = 3.732$ 1st Quadrant $2x = 74.9^0$and $x = 37.5^0$ $2x = 360 + 74.9^0 = 217.45^0$ 3rd Quadrant $x= 127.5^0$ Generally, $x = 37.5 + \pi k$ and $x = 127.5 + {}^{\pi}k$ where $\pi = 180$ and k = constant =1,2,3,4, 5 etc37.5 +180(1) =217.5^{0}127.5 + 180(1) = 307.5^0

b) $y = 3\sin(2x +30)$ $180^0 = 360^0 \div 2$ (2 is the coefficient of x). The period is the time (x axis - degrees) for the wave (curve) to complete one full cycle. Point X to Y corresponds to 0^0 to 180^0.

$y = 3\sin 2x$ Period $= 180^0$.

a. Amplitude refers to the range of the trigonometric graph $-3 \leq y \leq +3$.

b. Frequency refers to the number of complete curves in 360^0.

360^0 is the period of a $y = \sin x$ graph. For both graphs frequency = 2.

a. Consider the two points on both curves on the x axes: x,y = (75,0) and (90,0) , Phase Shift $= 90 - 75 = 15^0$. Likewise on the line y =3, x,y = (30,3) and (45,3), phase shift $= 45 - 30 = 15^0$

Lagging Curve : $y = 3\sin 2x$ Leading Curve : $y = 3\sin(2x +30)$

a. Domain of both graphs $= 0 \leq x \leq 360$

Example 2 Solve graphically the two simultaneous trigonometric equations:

$y = 4\cos4(x +30)$ and $y = 4\cos4x$ for the domain $0 \le x \le 360$

a. State the period of both graphs.
b. State the amplitude of both graphs.
c. State the frequency of both graphs.
d. State the phase shift, leading and lagging curve.
e. Domain of both graphs.

Table 1.19 $y = 4\cos4(x +30)$

x	0	15	30	37.5	45	60	75
y	-2	-4	-2	0	2	4	2

X	82.5	90	105	120	127.5	135	150
y	0	-2	-4	-2	0	2	4

x	165	172.5	180	195	210	217.5	225
y	2	0	-2	-4	-2	0	2

x	240	255	262.5	270	285	300	307.5
y	4	2	0	-2	-4	-2	0

x	315	330	345	352.5	360
y	2	4	2	0	-2

When $y = 0$, $4\cos4(x+30) = 0$ Applying compound angles in order to expand (refer to Chapter 4) $4(\cos (4x +120)) = 4(\cos4x\cos120 - \sin120\sin4x) = 4(-\frac{1}{2}\cos4x - \frac{\sqrt{3}}{2}\sin4x) = 0$ Solving we have $\tan4x = -\frac{1}{\sqrt{3}}$ $4x = -30^0$ Tan is –ve in the 2nd Quadrant $4x = 180^0 - 30^0 = 150^0$ and $x = 37.5^0$ Tan is negative in the 4th Quadrant $4x = 360^0 - 30^0 = 330^0$ $x = 82.5^0$ this curve intersects the x axis at intervals of 45^0. $82.5^0 - 37.5^0 = 45^0$ $127.5^0 - 82.5^0 = 45^0$ $172.5^0 - 127.5^0 = 45^0$ etc. $y = 4\cos4x$:

when y = 0: 0 = 4 cos 4x 4x = $\cos^{-1} 0 = 270^0$ Therefore, x = 67.5^0 this curve intersects the x axis at intervals of 45^0. $67.5^0 - 22.5^0 = 45^0 112.5^0 - 67.5^0 = 45^0 157.5^0 - 112.5^0 = 45^0$ etc.

Table 1.20 y =4cos4x

x	0	22.5	30	45	60	67.5	75
y	4	0	-2	-4	-2	0	2

x	90	105	112.5	120	135	150	157.5
y	4	2	0	-2	-4	-2	0

x	165	180	195	202.5	210	225	240
y	2	4	2	0	-2	-4	-2

x	247.5	270	285	292.5	300	315	330
y	0	4	2	0	-2	-4	-2

x	345	360
y	2	4

Refer to Fig 1.11.4. The points A, B, C, D and E are the y coordinates 4, -4, 0, 2 and -2. Reading off the intersection of the curves and comparing same values of x and y on both tables we have: x, y = 30, -2; 75, 2 120,-2; 165, 2 210, -2; 255, 2 270,-2; 300,-2 330, -2; 345,2.

For the point x, y = (345, 2): y= 4cos4 (345 +30) = 2 and y =4cos 4(345) = 2.

a. y= 4cos4(x +30) period = 90 360 ÷4 = 90 4 is coefficient of x and 360 = 1 revolution

y =4cos 4x period = 90 360 ÷4 = 90 4 is coefficient of x and 360 = 1 revolution

a. y= 4cos4(x +30) amplitude -4 ≤ y ≤ +4

y =4cos 4x amplitude -4 ≤ y ≤ +4

a. $y = 4\cos4(x + 30)$ frequency $= 4$ 4 complete cycles in 360^0

$y = 4\cos 4x$ frequency $= 4$ 4 complete cyles in 360^0

a.   Consider the two points from both tables at $y = 0$

$y = 4\cos4(x + 30)$; x, y $= 0, 37.5$

$y = 4\cos 4x$; x,y $= 0, 67.5$

the phase angle $= (67.5 - 37.5) = 30^0$ (Leading) $y = 4\cos 4x$ Lagging $4\cos4(x + 30)$ Leading

a.   Domain $= -\infty \le y \le +\infty$

Example 3 Sketch the curve of $y = \sec x$ for the domain $0 \le x \le 360$. Firstly, complete a table of values of $y = \sec x$ intervals of 30^0 (Table 1.21). The main purpose of this example to explain the following concepts: a) Domain : $x \ne 270^0$ $0^0 - 90^0$: $x \ne 90^0$; $0 \le x < 90$ $90^0 - 270^0$: $x \ne 90^0$ and $x \ne 270^0$ $90 < x < 270$

$270^0 - 360^0$: $x \ne 270^0$; $270 < x \le 360$

The curve is discontinuous at $x = 90^0$ and 270^0. These lines are vertical asymptotes. These x values are error values or undefined values when substituted into the initial equation.

b) Range : $1 \le y \le +\infty$ for $0 \le x < 90$ $x \ne 90^0$ and $x \ne 270^0$; y is not less than 1 For the interval $90 \le x < 270$ $-\infty \le y \le -1$.

y is not greater than -1.

Table 1.21

x 0   30   60   90   120   150   180
y 1   1.155 2   error -2   -1.155 -1
x 210   240   270 300 330   360
y -1.155 -2   error 2   1.155 1

It is very important to observe the fact that a value of x that gives an error or undefined value is a line of discontinuity. This line of discontinuity can also be referred to as an asymptote. From the table 1.21 above, the x values of 90 and 270 are lines of discontinuity.

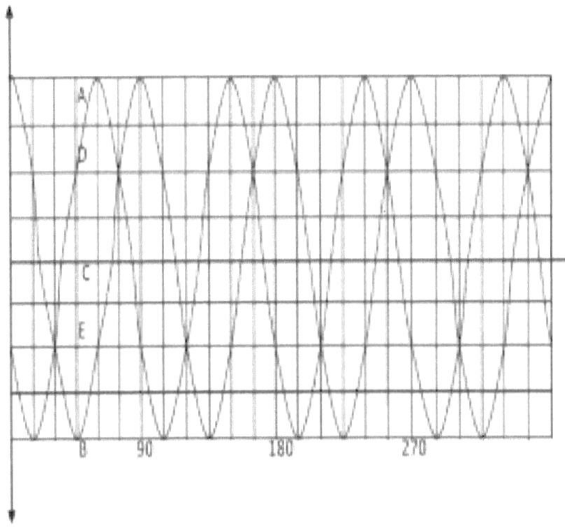

Fig 1.29 Graph of y = 4cos4x and y = 4cos 4(x + 30)

1.11.1 Domain and Range of a Trigonometric Function

The domain and range of a given trigonometric function is based on a discontinuous function or a continuous function.

The trigonometric function, y = sinx, is a continuous function. This is not the case for the function, y=secx. This function is a discontinuous function. Refer to Fig. 1.30.

The range of y = sinx is defined by the lines y = 1 and y = -1. Mathematically, range : -1 ≤ y ≤ +1 domain: -∞ ≤ x ≤ +∞

1.11.2 The inverse of a Trigonometric Function The tables and graph of y = sinx and x = siny are as shown. (See table 1.22, table 1.23 and Fig. 1.22)

Table 1.22 y = sinx

x	0	30	60	90	120	150	180
y	0	0.5	0.866	1	0.866	0.5	0

x	210	240	270	300	330	360
y	-0.5	-0.866	-1	-0.866	-0.5	0

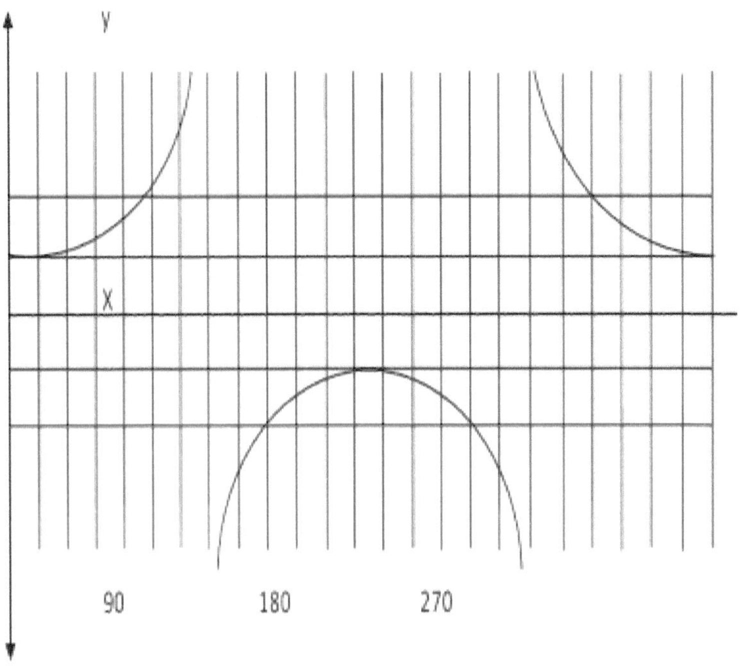

Fig. 1.30 Graph of y = sec x

y 0   30   60   90   120   150 180

x 0   0.5   0.866 1   0.866 0.5 0

y 210 240   270   300   330   360

x -0.5 -0.866 -1   -0.866 -0.5   0

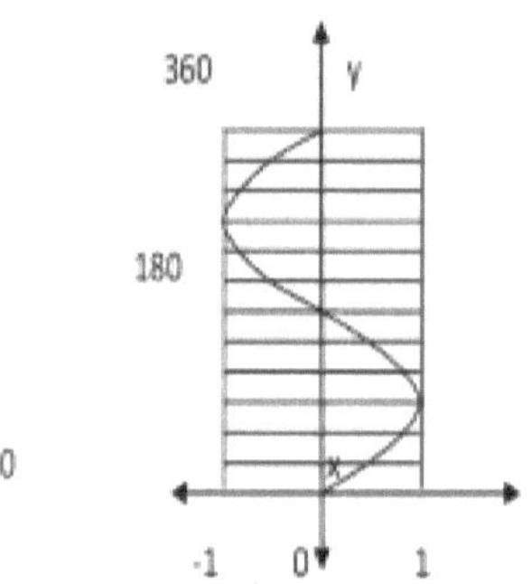

Fig 1.31 Graph of x = sin y

Exercise 1.6

1. Solve graphically the following set of equations for $0^0 \leq x \leq 360^0$.

Complete a table of x and y values and state your scale for both x and y axes.

a. y= 4sin(3x +45) and y = 4sin2x
b. y= 3sin x and y = 2cosx
c. y= 4sin(2x +45) and y = 4sin 2x

1. For each of the above set of equations a) – c):

a. State the period of both graphs.
b. State the amplitude of both graphs.
c. State the frequency of both graphs.
d. State the phase shift, leading and lagging curve.
e. Domain of the graphs.

1.12 Cubic Functions

The standard form of a cubic equation is $y = ax^3 + bx^2 + cx + d$. A cubic curve is a curve that has three x intercepts. **Example 1**

Draw to scale the cubic curve $y = x^3 - 9x^2 + 26x - 24$ by completing a table of values for x and y. Read off the y intercept, turning points and roots of the curve from your graph.

Solution:

Table 1.24

X	0	2	2.423*	2.5	3	3.577*	3.6	4	5
y	-24	0	0.385*	0.375	0	-0.385*	-0.384	0	6

y-intercept: -24 (x = 0)

Turning points of the graph from Fig 1.23 are: x1, y1=3.577, -0.385 x2, y2 = 2.423, 0.385.

These two pairs of coordinates have been marked with asterisk in the table.

The above pair of turning points will be calculated by method of differentiation in Chapter 6.

Roots: y = 0, x = 2, x = 3 and x = 4.

The roots of the cubic equation can be confirmed by applying the remainder theorem.

f(2) = 0; f(3) = 0 and f(4) = 0.

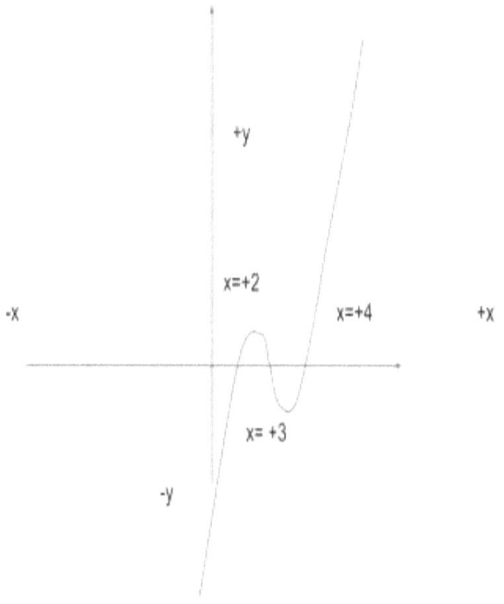

Fig. 1.32 Cubic Graph of $y = x^3 - 9x^2 + 26x - 24$

1.12.1 Domain and Range of a Cubic Function The domain and range of cubic curves are as follows. domain: $-\infty \leq x \leq +\infty$ range: $-\infty \leq y \leq +\infty$

1.12.2 Classification of Cubic Functions The Mathematics N4 National Exams cubic function problems are functions having different x and y values for the maximum, point of inflexion and minimum points. Refer to example .... (x1, y1, x2, y2 and x3 , y3).

A steady progression from the cubic functions $y=x^3$, $y=(x+3)^3$, $y=(x+3)^3$, $y=(x+3)^3$ and $y=(x+3)^3$ to the aforementioned cubic functions or curves is very fundamental. These functions are unique because the maximum points, points of inflexion and the minimum points intersect at the same x and y values. The first and second derivatives of $y=x^3$, $y=(x+3)^3$, $y=(x+3)^3$, $y=(x+3)^3$ and $y=(x+3)^3$ are one x value.

Another criterion for classifying cubic functions is based on the zero value of any of the coefficients of the expanded cubic function $y= ax^3+ bx^2+cx + d$.

For the cubic functions $y=x^3$, the values of b, c and d are zero. The expanded cubic functions of $y=x(x^2-9)$, $y=x(x^2-6)$, $y=x^2(x-5)$ and $y=x(x+4)^2$ are $y= x^3-9x$, $y=x^3-6x$, $y=x^3-5\,x^2$ and $y=x^3+8x^2 +16x$ respectively. The cubic function $y= x^3-9x$ has the coefficients of b and d equal to zero.

Thirdly, the concept of repeated root is evident in the cubic functions $y=x(x+4)^2$, $y=x(2x+5)^2$, $y=x^2(x-5)$. The repeated roots or x intercepts are x=-4, x=-4, x=-2.5, x=-2.5 and x= 0, x=0 respectively.

The fourth method of classifying cubic functions is the nature of the x intercepts or roots in terms of real values or unreal values. The cubic functions $y=(x+1)(x-2)(x-3)$ and $y=(x+3)(x+4)(x+5)$ have real roots or x intercepts. This is not the case for the cubic functions $y=(2x+7)(3x+1)(7x+1)$ and $y=(5x+7)(9x+1)(8x+1)$.

Finally, the positive or negative coefficient of the x^3 term determines the maximum point and minimum point as we progress in plotting the curve from the negative x axis to the positive x axis. Consider plotting the cubic functions having the positive coefficient of the x^3 term $y=x(x+4)^2$, $y=x(2x+5)^2$, $y=x^2(x-5)$ and the set of cubic functions $y=-(x(x+4)^2)$, $y=-(x(2x+5)^2)$, $y=-(x^2(x-5))$.

1.12.3 Inverse of a Cubic Function Based on the above example, the curve $x = y^3 - 9y^2 + 26y - 24$ is the inverse of $y = x^3 - 9x^2 + 26x - 24$. The inverse of the aforementioned cubic function is a very good exercise for students. The first step would be completing a table of values by interchanging the x and y values of the previous table. This exercise is an assessment of the students' understanding of the inverse of a cubic function.

Exercise 1.7

Determine graphically the turning points, y intercepts and the roots of the following cubic curves by a) completing a table of values for x and y.

1. $y = x^3 + 3x^2 - 16x - 48$
2. $y = 2x^3 + 7x^2 + 7x + 2$
3. $y = x^3 + 5x^2 - x - 5$
4. $y = x^3 - 2x^2 - 15x + 36$
5. $y = x^3 + 6x^2 + 11x + 6$

Chapter 2 Equations

2.1 Linear Equations A linear equation is the equation of a straight line. The standard form: y = mx + c m = slope or gradient of the line and c = y intercept. **2.1.1 Simultaneous Equations** There are two categories of Simultaneous Equations namely: a) Two linear simultaneous equations b) Simultaneous Equations: one linear and the other quadratic. c) Simultaneous Equations: one linear and the other is a circle.

In the first category, a very logical and practical approach to understanding Simultaneous Linear Equations is plotting two lines on the same axes. Firstly, two tables of values for both lines are completed and the lines are plotted based on these two points. The intersection of these two lines gives us the x and y values of the graphical solution to the problem.

Likewise, for the second category, the x and y values of two points of intersection of the straight line and the parabola are read off in order to obtain the required coordinates.

Examples of graphical solution of simultaneous equations can be found in Engineering Mathematic N3 series. Simultaneous linear equations consist of two unknowns (x and y) and two linear equations as shown in the examples 1 and 2 below.

2.1.2 Word Problems Word Problems are problems that are based on any one of the following categories of equations as stated: a) Two linear simultaneous equations b) Simultaneous Equations: one linear and the other quadratic. c) One linear equation and an equation of a circle. d) One linear equation and a rectangular hyperbola e) One quadratic equation only.

They are called Word Problems because students are required to firstly define or state the x and y variables based on the Word Problem.

The second step is writing the two equations in x and y based on the details as provided in the problem. Examples 1-4 clearly explains the procedures applied in solving simultaneous equations.

Example 1 The difference in age between Jacob and Jabulani is 20 years. Five years ago, Jacob's age was twice Jabulani's age. Calculate Jacob's age and Jabulani's age. Let x = Jacob's age and y = Jabulani's age. Jacob is older than Jabulani. x – y = 20 (1) x -5 = 2(y-5) (2) Five years ago From (1), x = 20 + y substituting this into (2): (20 + y) – 5 = 2y – 10 y = 25 years and x = 20 + 25 = 45 years.

The two equations above are two linear equations. This example has been solved by applying the method of substitution.

Example 2 Calculate the volume in milliliters of each of 10% nitric acid and 15% nitric acid that will be mixed in order to obtain 150ml of a solution containing 11.67% nitric acid. Let x = 10% of unknown volume and y = 15% of unknown volume x +y = 150 ml (1) 10x +15y = 150ml X 11.67 (2) Equation (1) X 10 : 10x + 10y = 1500 (3) Subtract (3) from (2) (10x +15y) – (10x + 10y) = 1750 – 1500 5y = 250 y = 100ml x= (150 – 100) = 50ml

This example has been solved by applying the method of elimination.

Example 3. Two men, A and B, start walking in the same direction from the same spot. After two hours A is 25km ahead of B and A is at a distance of 125km from the starting point. Calculate the speed of A and B. Let speed of A = x km/h and speed of B = y km/h

Distance equation: 2hours(x) – 2hours(y) = 25km (1) 2hours(x) + 2hours(y) = (125 +100) km (2) Subtract Equ. 2 from Equ. 1 -4y = -200 y = 50km/h and 2x – 2(50) = 25 x = 62.5km/h

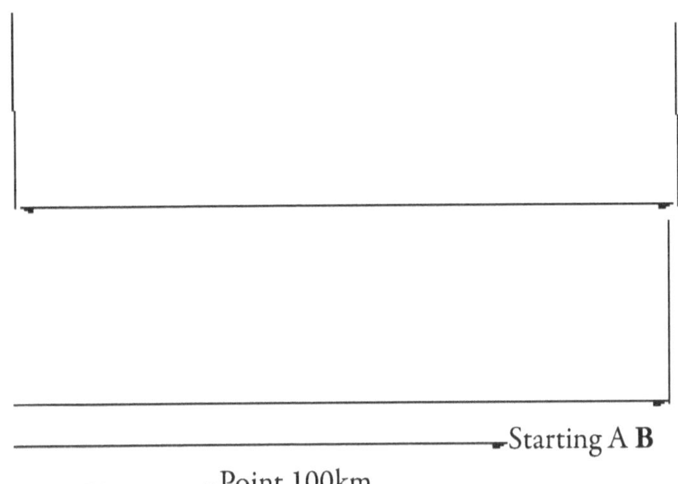

Fig. 2.1 Distance covered by two men after 2 hours

Example 4 A rectangular box has a diagonal of 1m. The perimeter of the box was measured and recorded as 2.8m. Find the length and breadth of the box. Let the length = x metres and breadth = y metres. Perimeter : $2x + 2y = 2.8$metres (1) $x + y = 1.4$ and $x = 1.4 - y$ (3)

Pythagoras Theorem: $x^2 + y^2 = (1m)^2$ (2) Substituting (3) into (2) $(1.4 - y)^2 + y^2 = 1$ $1.96 - 2.8y + y^2 + y^2 = 1$ $2y^2 - 2.8y + 0.96 = 0$ Applying the Quadratic Formula

$$y = \frac{-(-2.8) \pm \sqrt{(-2.8)^2 - 4(2)(0.96)}}{2(2)}$$

$= 0.8$m ; 0.6m Substitute 0.8m into (3) $x = 1.4 - 0.8 = 0.6$m

The above example illustrates simultaneous equations: one linear and the other a circle. There are two unknowns in the equations.

2.2 Simultaneous Equations (Three Equations and Three Unknowns)

There are three variables (x, y and z) and three equations as against two in the previous examples. The steps required to solve these equations are stated thus: Using Example 6 below as illustration: a) Eliminate x from Equ. (1) and (2) in order to get Equ. 5

in y and z. b) Eliminate x from Equ. (1) and (3) so that we can obtain an equation in y and z (Equ. 7). c) Eliminate y from Equ. 5 and Equ. 7 in order to calculate the value of z. d) Substitute the value of z into Equ. 5 or Equ. 7 in order to solve for y. e) Substitute the values of y and z into any one of the three original given equations to calculate the value of x. f) We may confirm our answers by substituting the values of x, y and z into any one of the three equations.

Example 6 Solve for x, y and z given the following equations:

$x + z = 10$ (1) $3x + y - 3z = 15$ (2) $5x - y - 6z = -5$ (3) Solution: Equation (1) X 3: $3x + 3(0)y + 3z = 3(10)$ $3x + (0)y + 3z = 30$ (4) Subtract (2) from (4) $3x + (0)y + 3z = 30$ (4) $3x + y - 3z = 15$ (2) $(3x + (0)y + 3z) - (3x + y - 3z) = 30 - 15$ $-y + 6z = 15$ (5) Equation (1) X 5 $5x + 5(0)y + 5z = 5(10)$ $5x + 5(0)y + 5z = 50$ (6) Subtract Equ. 6 from Equ. 3 $(5x - y - 6z) - (5x + 5(0)y + 5z) = -5 - 50$ $-y - 11z = -55$ (7) Subtract (7) from (5) We now have two unknowns and two equations as follows:

$-y + 6z = 15$ (5) $-y - 11z = -55$ (7) $17z = 70$ $z = 4.118$ Alternatively, $(-y + 6z) - (-y - 11z) = 15 - - 55$ $17z = 70$ $z = 4.118$ Substitute z into (5), we have $y = 6(4.118) - 15 = 9.706$ Substitute z into (1) $x = 10 - 4.118 = 5.882$ Confirmation: $3(5.882) + 9.706 - 3(4.118) = 15$ This method is known as the method of elimination. The x term is eliminated in order to obtain two equations in terms of y and z. Likewise, the y term was eliminated so that the value of z can be calculated. Usually, the elimination of a variable is carried out in column form as stated above. However, most students find this method difficult to understand. The calculation: $(-y + 6z) - (-y - 11z) = 15 - - 55$ and the column elimination method: $-y + 6z = 15$ (5) $-y - 11z = -55$ (7) $17z = 70$ $z = 4.118$ will give the same answer.

2.3 Quadratic Equations Quadratic equations have the standard form $y = ax^2 + bx + c$. The following examples explain the methods that are applied to solve quadratic equations.

Example 7 Find the values of x for the equation $6x^2 + 17x + 7 = 0$ by applying the following methods:

a. Factorization
b. Quadratic Formula
c. Completing the square
d. Graphical method

Solution:

a. $6x^2 + 17x + 7 = 0$

Table 2.1

Product	Sum (x terms)
$6 \times 7 = 42$	$6x + 7x = 13x$
$42 = 3 \times 14$	$3x + 14x = 17x$
$42 = 2 \times 21$	$2x + 21x = 23x$

Replace middle term 17x with 3x + 14x and then factorize:

$6x^2 + 3x + 14x + 7 = 0$ $(6x^2 + 3x) + (14x + 7) = 0$ $3x(2x + 1) + 7(2x + 1)$ $(2x + 1)(3x + 7) = 0$

$2x + 1 = 0$ $3x + 7 = 0$ Therefore, $x = -0.5$ and $x = -2.33$

a. Compare $6x^2 + 17x + 7 = 0$ with the standard equation $ax^2 + bx + c = y$.

$a = 6, b = 17$ and $c = 7$

$$x = \frac{-(17) \pm \sqrt{(17)^2 - 4(6)(7)}}{2(6)}$$ $x_1 = -0.5$ and $x_2 = -2.33$

a. $6x^2 + 17x + 7 = 0$ $6x^2 + 17x = -7$ divide both sides by 6

$6x^2$ is not a perfect square x^2 is a perfect square

$X^2 + \dfrac{17x}{6} = \dfrac{-7}{6}$ Add the square of half the coefficient of x to both sides

$$X^2 + \frac{17x}{6} + \left[\frac{17}{6} \times \frac{1}{2}\right]^2 = \frac{-7}{6} + \left[\frac{17}{6} \times \frac{1}{2}\right]^2$$

$\left[X + \dfrac{17}{12}\right]^2 = \dfrac{121}{144}$ Find square root of both sides

$X + \dfrac{17}{12} = \pm \dfrac{11}{12}$ $x_1 = -0.5$; $x_2 = -2.33$

a. Choose a suitable scale and draw the graph of $y = 6x^2 + 17x + 7$ based on the table below.

Table 2.2

x	-3	-2.33	-2	-0.5	0
y	10	0	-3	0	7

Solution: read off the x values of the x intercepts from your graph.

Example 8 Find the values of x for the equation $14x^2 + 23x + 3 = 0$ by applying the following methods:

a. Factorization
b. Quadratic Formula
c. Completing the square

Solution:

a. $14x^2 + 23x + 3 = 0$

Table 2.3

Product	Sum (x terms)
$14 \times 3 = 42$	$6x + 7x = 13x$
$42 = 3 \times 14$	$3x + 14x = 17x$
$42 = 2 \times 21$	$2x + 21x = 23x$

Replace middle term 23x with 2x + 21x and then factorize:

$$14x^2 + 2x + 21x + 3 = 0$$

$$2x(7x + 1) + 3(7x + 1) = 0 \quad (7x + 1)(2x + 3) = 0 \quad 7x + 1 = 0$$

$$2x + 3 = 0 \quad x = -\frac{1}{7} \text{ and } x = -\frac{3}{2}$$

a. $x = \dfrac{-(23) \pm \sqrt{(23)^2 - 4(14)(3)}}{2(14)} \qquad x_1 = -\frac{1}{7} \quad x_2 = -\frac{3}{2}$

b. $14x^2 + 23x = -3 \quad x^2 + \frac{23}{14}x = -\frac{3}{14}$

$$X^2 + \frac{23}{14}x + \left[\frac{23}{14} \times \frac{1}{2}\right]2 = -\frac{3}{14} + \left[\frac{23}{14} \times \frac{1}{2}\right]2$$

$$\left[x + \frac{23}{28}\right]2 = \frac{361}{784} \quad x + \frac{23}{28} = \pm\left[\frac{361}{784}\right]0.5 \quad x = \pm\frac{19}{28} - \frac{23}{28} \quad x_1 = -\frac{4}{28} = -\frac{1}{7} \text{ and } x_2 = -\frac{42}{28} = -\frac{3}{2}$$

Example 9 Evaluate the values of x for the equation $10x^2 - 19x + 6 = 0$ by applying the following methods:

a. Factorization
b. Quadratic Formula
c. Completing the square

Solution:

a. $10x^2 - 19x + 6 = 0$

Table 2.4

Product	Sum (x terms)
$10 \times 6 = 60$	$6x + 10x = 16x$
$60 = 4 \times 15$	$4x + 15x = 19x$
$60 = -4 \times -15$	$-4x - 15x = -19x$
$60 = 3 \times 20$	

Replace middle term -19x with -4x - 15x and then factorize:

$10x^2 - 4x - 15x + 6 = 0(10x^2 - 4x) - (15x - 6) = 0$ observe the sign change in second bracket

$2x(5x - 2) - 3(5x - 2) = 0$

$(5x - 2)(2x - 3) = 0 \quad x_1 = \dfrac{2}{5} \quad x_2 = \dfrac{3}{2}$

a. $x = \dfrac{-(-19) \pm \sqrt{(-19)^2 - 4(10)(6)}}{2(10)}$ x1 = 0.4   x2 = - 1.5

b. $10x^2 - 19x = -6$ Divide both sides by 10

$$X^2 - \frac{19}{10}x = -\frac{6}{10}$$

$$X^2 - \frac{19}{10}x + \left[\frac{19}{10}x\frac{1}{2}\right]2 = -\frac{6}{10} + \left[\frac{19}{10}x\frac{1}{2}\right]2$$

$\left[x - \dfrac{19}{20}\right]2 = \dfrac{121}{400}$ Solving for x, we have, x1 = $\dfrac{2}{5}$ x2 = $\dfrac{3}{2}$

Example 10 A truck driver travelling at a constant speed covers a distance of 300km. If he increases his speed by 15km/h, he will arrive at his destination 40 minutes earlier. Determine his speed.

Applying the formula, time = $\dfrac{\text{distance}}{\text{speed}}$ and considering the fact that

x = constant speed, $\dfrac{300}{x} - \dfrac{300}{x+15} = \dfrac{2}{3}$hrs, multiplying both sides by $3x(x +15)$, we have

$300[3(x+15) - 3x] = 2x(x+15)$ Simplifying and applying the Quadratic Formula: $150[3x + 45 - 3x] = x(x + 15)$ $150(45) = x(x + 15)$ $6750 = x^2 + 15x$ $x^2 + 15x - 6750 = 0$

x= 75km/h x= -90km/hr (choose 75km/h since it is positive)

Exercise 2.1 Solve for x and y: 1. 3x + y = 15 . 6x – 3y = 4 2. 7x +2y +10 = 0 . x+3y -2=0 3. 10y = 2x -1 . 3y= 6x -10 4. y=x -10 . y=4x+20 Solve for x, y and z 5. 2x +6y +z+20=0 . 3x+y – z -1 = 0 . x+5y +6z – 10 =0 6. x-z =10 . 3x +7y + 8z = 12 . 4x+2y+9z =3 7. 4z + 6y –x =5 . x-y +8z = -2 . 10z -3x -7y = -1 8. 2z +3y +7x +9 =3 . 5x -7y +3z -2c ****

. 2z +3x -2y +4= -4 Solve for x by factorizing: 9. $21x^2 – 46x – 7 = 0$ 10. $4x^2 – 20x + 21 = 0$ 11. $5x^2 + 13x – 18 = 0$ 12. $5x^2 – x - 18 = 0$ 13. $6x^2 + 17x + 5 = 0$

Solve for x by applying the quadratic formula: 14. $21x^2 - 46x - 7 = 0$ 15. $4x^2 - 20x + 21 = 0$ 16. $5x^2 + 13x - 18 = 0$ 17. $5x^2 - x - 18 = 0$ 18. $6x^2 + 17x + 5 = 0$ Solve for x by completing the square: 19. $-46x + 21x^2 = 7$ 20. $21 = -4x^2 + 20$ 21. $5x^2 + 13x - 18 = 0$ 22. $-x - 18 = -5x^2$ 23. $6x^2 + 17x + 5 = 0$ 24. 5 litres of petrol and 10 litres of paraffin costs R181.00 while 15 litres of petrol and 6 litres of paraffin cost R243.00. Calculate the price of petrol and paraffin. 25. Pretoria Central Business District (cbd) is 150km from Johannesburg Cbd. While driving from Pretoria to Johannesburg it takes the driver 45 minutes and on the return trip it takes him 1hour. Assuming that there is no time wasted upon arriving in Pretoria, calculate the speed of the driver on both trips (Pretoria – Johannesburg and Johannesburg – Pretoria). 26. While travelling downstream a boat moves a distance of 45km in 2 hours and it moves a distance of 15km in 3 hours if it travels upstream. What is the speed of the boat when travelling upstream and downstream? (Hint: Let x = boat speed; y = stream speed) 27. Two numbers are in the ratio 4:5 and their difference is 30. Find the numbers. 28. The base of a right-angled triangle is 12m more than its height. The area of the right-angled triangle is $32m^2$. Find the height and base of the triangle. 24. The base of a right-angled triangle is 12m more than its height. The area of the right-angled triangle is $32m^2$. Find the height and base of the triangle. 25. A motorist has to travel a total distance of 250km. if he increases his average speed by 25km/hr, he will arrive his destination 45minutes earlier. Calculate the average speed. 26. Two consecutive even numbers are squared and summed up to get 8452. What are the numbers? 27. 75% of the difference of two numbers is 12. 50% of the sum of the two numbers is 48. Find the two numbers. 28. The length of a rectangle is 12m more than the breadth. The area of the rectangle is $108m^2$. Calculate the breadth and length of the rectangle. 29. The hypotenuse of a right-angled triangle is 8cm longer than the shortest side of the two right angled sides. Calculate the lengths of the 3 sides given that the shortest side is 12cm.

2.4 Exponential and Logarithmic Equations The table below explains the relationship between an exponential equation and its corresponding logarithmic equation.

Table 2.5

1.	$2^4 = 16$	$y= \log_2 16 = \log_2 2^4 = 4$
2.	$3^6 = 729$	$y= \log_3 729 = \log_3 3^6 = 6$
3.	$5^3 = 125$	$y= \log_5 125 = \log_5 5^3 = 3$
4.	$8^4 = 4096$	$y= \log_8 4096 = \log_8 8^4 = 4$
5.	$10^5 = 100\ 000$	$y= \log_{10} 100\ 000 = \log_{10} 10^5 = 5$

$y= \log_2 16 = \log_2 2^4 = 4$ the base of this logarithmic equation is 2 because $2^4 = 16$. $y= \text{Log}_2 16 = \log_2 2^4 = 4$ this equation is explained thus: the logarithm of 16 to the base 2 is equal to 4. For the equation, $2^4 = 16$, 4 is called an exponent. $2^4 = 16$ is pronounced 2 raised to the power 4.

2.4.1 Exponential Laws

1. $a^b \times a^c = a^{b+c}$ Example $8 \times 16 = 2^{3+4} = 2^7$
2. $a^b \div a^c = a^{b-c}$ Example $16 \div 8 = 2^{4-3} = 2$
3. $(a^b)^c = a^{b \times c}$ Example $(2^3)^2 = 2^{3 \times 2} = 2^6$
4. $A^0 = 1$

Example 1: $3^x = 729$ $729 = 3^6$ $3^x = 3^6$ Therefore, $x = 6$

Example 2: $4^{x+3} = 256$ $256 = 4^4$ $4^{x+3} = 4^4$
Equating Exponents: $x + 3 = 4$ Solving for x: $x = 1$

Example 3: $9^x \times 9 = 6561$

$6561 = 9^4$ Writing in terms of Powers of 9: $9^{X+1} = 9^4$ Equating exponents: $x + 1 = 4$ $x = 3$

Example 4: $[\frac{1}{2}]^x = 64$

$[\frac{1}{2}]^x = 64$ $64 = [2^3]^2$ $[\frac{1}{2}]^x = [2^{-1}]^x$

$[2^3]^2 = [2^{-1}]^x$ $2^6 = 2^{-x}$
Equating Exponents: $6 = -x$ and $x = -6$

Example 5: $2^{(x+2)(x+2)} = 1$

$2^{(x+2)(x+2)} = 2^0$ since $2^0 = 1$

$(x+2)(x+2) = 0$ $x+2 = 0$ and $x = -2$

Example 6: $3^{3x+1} + 3^{3x+1} = 162$

$3^{3x}.3 + 3^{3x}.3 = 2 \times (3^2)^2$

$3(3^{3x} + 3^{3x}) = 2 \times 3^4$

$2 \times 3(3^{3x}) = 2 \times 3^4$

$3^{3x+1} = 3^4$ Equating Exponents: $3x + 1 = 4$ and $x = 1$

Example 7 Calculate the values of x and y given that: $8^{2x+y} = 4^{y+4}$ and $x = 4y$ Express 8 and 4 in terms of powers of 2 in the first equation. $2^{3(2x+y)} = 2^{2(y+4)}$ and $2^{6x+3y} = 2^{2y+8}$ Equating exponents: $6x + 3y = 2y + 8$ $y = 8 - 6x$ (1) $x = 4y$ (2) Solving simultaneously: $y = 8 - 6(4y)$ $y = 0.32$ $x = 1.28$

2.4.2 Logarithmic laws. a) $\text{Log}_c a + \text{Log}_c b = \log_c (ab)$ Example $\log_{10} 25 + \log_{10} 4 = \log_{10} (25 \times 4) = 2$ b) $\text{Log}_c a - \text{Log}_c b = \log_c (a/b)$ Example $\log_{10} 100 - \log_{10} 10 = \log_{10} (100/10) = 1$ c) $\text{Log}_c a = [\log_d a]/[\log_d c]$ Example $\log_4 64 = [\log_2 64]/[\log_2 4] = 3$ The third law is known as the Change of base Formula

2.4.3 Logarithmic Equations Example 1 Given that $\log 100 + \log x = 5$, solve for x. Usually, when the base is not stated, the base is 10. $\text{Log}(100x) = 5$ (first Law) $10^5 = 100x$ (Conversion to Exponential Form) $x = 1000$ Alternatively , $\log 10^2 + \log x = 5$ $2\log 10 + \log x = 5$ $\log x = 5 - 2 = 3$ convert $\log x = 3$ into exponential form, we have, $10^3 = x$

Example 2 Calculate the value of x for the equation: $\log 2x + \log(3x + 1) = 2$ $\log(2x(3x+1)) = 2$ Simplifying: $\log (6x^2 + 2x) = 2$ $6x^2 + 2x = 10^2$ Exponential Form $6x^2 + 2x - 100 = 0$ $x = \dfrac{-2 \pm \sqrt{2^2 - 4(6)(-100)}}{2(6)}$ $x_1 = 3.92$; $x_2 = -4.25$ (not applicable)

Example 3 Solve for x given the equation: $\log(2x-3) + \log(x+1) = 2\log x$ $\log(2x-3)(x+1) = \log x^2$ expanding and comparing left and right hand sides $\log[2x(x+1) - 3(x+1)] = \log x^2$ $2x^2 - x - 3 = x^2$ simplifying: $x^2 - x - 3 = 0$ $x = 2.303$ $x = -1.03$ (not applicable)

Example 4 $x = [30^{\log 6}]/[6^{\log 3}]$ Taking log of both sides and simplifying $\log x = \log [[30^{\log 6}]/[6^{\log 3}]]$ Taking log of both sides and simplifying $\log[30^{0.7782}]/[6^{0.4771}] = \log 6$ $\log x = \log 6$ $x = 6$

Alternatively, logx =log $30^{\log 6}$ – $\log 6^{\log 3}$ log x =(log6)(log 30) – (log3)(log6) log x = 0.7782 conversion into exponential form: x = $10^{0.7782}$ = 6

Example 5 Evaluate x given that log 4096^x = 4 xlog 4096 = 4 Therefore, x = 4/(log 4096) x= 1.1073

Example 6 Solve for x: log$_x$ 81 = log$_{11}$ 121 log$_x$ 9^2 = log$_{11}$ 11^2 Therefore 2log$_x$ 9 = 2 log$_{11}$ 11 x = 9

Exercise 2.2 Solve for x: 1. x^5=16 2. $3x^{1.5}$= 81 3. $x^{1/2}$ = 121 4. $X^{3/2}$= 9 5. $X^{4/3}$= 81 6. $100^{2x + 2}$ = $1000^{x + 4}$7.$(0.16)^{x +2}$ = 64^{2x}8. $9^{x - 2}$ = $(1/3)^x$9. 8^x = $(0.015625)^{x +4}$10. log $(0.4)^x$ = 16 11. log(2x

-3) + 2 = 0 12. log (2x +1) + log(x-3) = logx 13. $\dfrac{\dfrac{\log\log 15}{\log 4}}{\dfrac{\log x}{\log 3}}$ =

14. Log x =$\dfrac{\log 4}{\log 16}$ 15. Log 16 + log x = 2 16. X = $\dfrac{100^{\log 15}}{4^{\log 3}}$

17. Log $6x^4$ + logx^2 = log 9 – log x 18. Log 16^x = 3

1. Formula Calculations

Formulas are applied in design calculations. This section emphasizes the importance of making a letter or symbol the subject before substituting in order to solve for our final answer. It would be quite tedious to substitute the given data at the beginning of the calculation. The following example illustrates this.

Example 1 Make time, t, the subject of the formula: a) s = ut + 0.5 at^2 b) A = P(1 + 0.01r)^t a) s = ut + 0.5 at^2This is a quadratic equation in terms of t. The equation should be written thus: 0.5at^2 + ut – s = 0 0.5at^2 + ut =

+s This equation has been written in the quadratic formula standard form $y = ax^2 + bx + c$ where $y = 0$ Based on this, $a = 0.5a$, $b = u$ and $c = -s$. Substituting these values into our quadratic formula will give us the required answer. $t = \dfrac{-u \pm \sqrt{u^2 - 4(0.5a)(-s)}}{2(0.5a)}$ b) $(1 + 0.01r)^t = \dfrac{A}{P}$ Take log of both sides $\log(1 + 0.01r)^t = \log\left[\dfrac{A}{P}\right]$ Applying Law of Logarithm.

$t \log(1 + 0.01r) = \log\left[\dfrac{A}{P}\right]$ $t = \left[\log\left[\dfrac{A}{P}\right]\right] \div \left[\log(1 + 0.01r)\right]$

Example 2 Calculate the nth term of the geometric series given that: a) $T = ar^{n-1}$ $r = 0.5$ and $a = 0.5$ and $T = 0.03125$ b) $S = \dfrac{a(r^n - 1)}{r - 1}$ $r = 0.5$, $a = 0.5$ and $S = 0.96875$ a) $r^{n-1} = T/a$ Take log of both sides $\log r^{n-1} = \log(T/a)$ $(n - 1)$ $\log r = \log(T/a)$ $n - 1 = \dfrac{\log\log\left(\frac{T}{a}\right)}{\log\log r}$ and $n = \dfrac{\log\log\left(\frac{T}{a}\right)}{\log\log r} + 1$

Substituting, we have $n = \dfrac{\log\log\left(\frac{0.03125}{0.5}\right)}{\log\log 0.5} + 1 = 5^{\text{th}}$ term

b) $r^n - 1 = \dfrac{S(r-1)}{a}$ (add 1 to both sides) $r^n = \dfrac{S(r-1)}{a} + 1$ (take log of both sides) $\log r^n = \log\left[\dfrac{S(r-1)}{a} + 1\right]$ $n \log r = \log\left[\dfrac{S(r-1)}{a} + 1\right]$ Therefore: $n = = \left[\log\left[\dfrac{S(r-1)}{a} + 1\right]\right] \div \left[\log r\right]$

Substituting: $n = \left[\log\left[\dfrac{0.96875(0.5 - 1)}{0.5} + 1\right]\right] \div \left[\log 0.5\right] = 5^{\text{th}}$ term

Example 3 Given the formula $I = \dfrac{E}{[R^2 + X^2]^{0.5}}$, make X the subject of the formula. $[R^2 + X^2]^{0.5} = \dfrac{E}{I}$ Square both sides

$R^2 + X^2 = [\frac{E}{I}]^2$ $X^2 = [\frac{E}{I}]^2 - R^2$ Square root of both sides X

$= [[\frac{E}{I}]^2 - R^2]^{0.5}$ The power or exponent 0.5 is the same as writing square root sign.

Generally, in order to make any of the following the subject of the formula, it is necessary to take logarithm of both sides: a) the exponent or power b) the letter or symbol which has been raised to a certain power.

_____**Exercise 2.3** Make the letter or symbol in brackets the subject of the formula. 1. $f = \dfrac{E}{2\pi\sqrt{LC}}$ (c) 2. $I_{rms} =$ $I_{peak}(\phi/(2\pi))^{0.5}$ (ϕ)

/__________3. $f = \dfrac{1}{2\pi}\sqrt{\dfrac{1}{LC} - [\dfrac{R}{2L}]^2}$ (R) 4. $S = \dfrac{a(r^n - 1)}{r-1}$ (n) 5. $T_1/T_2 = e^{\mu\Theta}$ (μ) 6. $y = \log_8 x$ (x) 7. $v = (2gh)^{0.5}$ (h) 8. $A = P(1 + r)^n$ (r)

Chapter 3 Determinants

3.1 Introduction Consider the three equations and three unknowns of Example 6, Section 2.2 as stated below. This set of equations can be written in matrix form by firstly rewriting the first equation as stated below.

x +z = 10 (1) this is written x + 0(y) + z = 10
3x + y – 3z = 15 (2)
5x – y – 6z = -5 (3)

The above three equations can be written in a square matrix form or 3 X 3 matrix thus: 1 0 1 10

3 1 -3 = 15

5 -1 -6 -5

z column y column x column

3 X 3 matrix means 3 rows by 3 columns. The first number specifies the row while the second number specifies the column. Generally the above matrix is written as follows: A11 A12 A13

A21 A22 A23

A31 A32 A33

Where A31 = Element in the third row and in the first column.

A23 = Element in the second row and in the third column.

The minor of A21 is M21 = A12 A13

A32 A33

The minor of a matrix is obtained by crossing out the corresponding column and row in which that element is positioned.

The co factor of A21 is (-1) X M21

(+1)A11 (-1)A12 (+1)A13
(-1)A21 (+1)A22 (-1)A23
(+1)A31 (-1)A32 (+1)A33

Based on the above matrix above, the position of A21 has a negative sign hence the expression: (-1) M21 = - M21.

The cofactor of A21 , M21 = (-1) A12 A13

A32 A33

The relationship between a minor and a cofactor is that the cofactor has a positive or negative sign attached to it based on the position of the element in the matrix. The value of a determinant is illustrated in the following example below.

Example 1: Evaluate the determinant of the matrix above as follows: a) along the first row b) along the third column c) along the second row

Solution:

a. $(+1)(1)$ 1 -3 = [(1X-6) - (-3 X-1)] = -9[1 = A₁₁]

-1 -6

$(-1)(0)$ 3 -3 = 0[3(-6) - 5(-3)] = 0[0 = A₁₂]

5 -6 $(+1)(1)$ 3 1 = 1 [3(-1) - 5(1)] = -8 [1 = A₁₃]

5 -1

Determinant = -9 + 0 - 8 = -17

a. $(+1)(1)$ 3 1 + $(-1)(-3)$ 1 0 + $(+1)(-6)$ 1 0

5 -1 5 -1 3 1 = -8 – 3 -6 = -17

a. $(-1)(3)$ 0 1 + $(+1)(1)$ 1 1 + $(-1)(-3)$ 1 0

-1 -6 5 -6 5 -1

. 3 = A₂₁ 1= A₂₂ -3 = A₂₃

= -3 -11 -3 = -17 The arrows show the order in which the determinant is evaluated. The product of top left element and bottom right element minus the product of bottom left element and top right element. Generally, the determinant of a matrix is the same value regardless of the row or column along which it is calculated.

Example 2 Prove that $x = d_x/d_o$ and $y = d_y/d_o$ for the 2 X 2 matrix: $a_1 x + b_1 y = k_1$ $a_2 x + b_2 y = k_2$

The method of elimination will be applied in order to eliminate x and y. In order to eliminate x, the first equation is multiplied by a_2 while the second equation is multiplied by a_1. $a_1 a_2 x + b_1 a_2 y = k_1 a_2$ (1) $a_1 a_2 x + b_2 a_1 y = k_2 a_1$ (2)

Subtracting Equ. 2 from Equ. 1, we have, $b_1 a_2 y - b_2 a_1 y = k_1 a_2 - k_2 a_1$ factorizing : $y(b_1 a_2 - b_2 a_1) = k_1 a_2 - k_2 a_1$ Solve for y: $y = [k_1 a_2 - k_2 a_1] \div [b_1 a_2 - b_2 a_1]$ Equ. 1 Applying the determinants method on the original equation:

$$\left\| \begin{array}{cc} a_1 & b_1 \\ a_2 & b_2 \end{array} \right\| = \begin{array}{cc} k_1 \\ k_2 \end{array}$$

$$d_o = \left| \begin{array}{cc} a_1 & b_1 \\ a_2 & b_2 \end{array} \right| = a_1 b_2 - b_1 a_2$$

$$d_y = \left\| \begin{array}{cc} a_1 & k_1 \end{array} \right\| = a_1 k_2 - a_2 k_1$$

$a_2 \ k_2$

The determinant d_y is obtained by replacing the y column by the k column. This procedure is the equivalent of eliminating one of the variables in the simultaneous equation calculations. Now, by comparing $a_1k_2 - a_2k_1$ and $k_1 \ a_2 - k_2 \ a_1$, we can see that: $k_1 \ a_2 - k_2 \ a_1 = + [\ k_1 \ a_2 - k_2 \ a_1 \ ]$

$= - - [\ k_1 \ a_2 - k_2 \ a_1]$

$= k_2 \ a_1 - a_2k_1$ Similarly, $b_1 \ a_2 - b_2 \ a_1 = + [\ b_1 \ a_2 - b_2 \ a_1]$

$= - - [b_1 \ a_2 - b_2 \ a_1]$

$= b_2a_1 - b_1a_2$

Therefore, $y = d_y/d_0$ and $y = [a_1k_2 - a_2k_1] \div [b_2a_1 - b_1a_2]$ Likewise, Cramer's rule for $x = d_x/d_0$ may be proven.

This example shows the relationship between two equations and two unknown problems and the solution of 2 X 2 matrix by applying Cramer's rule.

3.2 Cramer's Rule : Solving 2 X 2 matrix

Example 3 Solve for x and y by applying Cramer's Rule x - y= 20 x- 2y = -5 1 -1 = 20 1 -2 -5

$d_0 = (-2) - 1(-1) = -2 +1 = -1$

$D_x = 20 -1 . -5 -2$

= 20(-2) – (-5)(-1) = -40 – 5 = -45

d_y = 1 20 = 1(-5) – 1(20)

1 -5 = -5 -20 = - 25 x= d_x/d_0 =- 45 /-1 = 45 y = d_x/d_0 = -25/-1 = 25

Example 4 Calculate x and y: 2x – 2y = 25 2x + 2y = 225

2 -2 = 25 2 2 = 225

d_x = 25 -2 = 25(2) –(225)(-2) = 500

225 2 d_y = 2 25 = 2(225) – 2(25) = 400

2 225

x= d_x/d_0 = 500 /8 = 62.5 y = d_x/d_0 = 400/8 = 50

Exercise 3.1 Solve for x and y: 1. x – y = 10 ; x -2y = 3

2. 2x -10 = 3y ; x – y= -6

3. y = 0.5x +3 ; y = 2.5x +10

4. 2y – 3x = -1; 3y – 6x = 10 5. 7x = 8y -3 ; 3x + 6y = 9

6. 10y= 2x -5 ; y = 4x – 6 7. y=2x + 3 ; y=5x -3 8. 7x =y +3; x= y-3

3.3 Cramer's Rule: Solving 3 X 3 matrix

Example 1. Solve for x, y and z given that:

$x + z = 10$ (1) 1 0 1 10
$3x + y - 3z = 15$ (2) 3 1 -3 = 15
$5x - y - 6z = -5$ (3) 5 -1 -6 -5

It is important to take note of the fact that the coefficient of y in the first equation is zero. The value of x, y and z can be calculated by applying the formula: $x = d_x/d_0$, $y = d_z/d_0$ and $z = d_z/d_0$

d_x = 10 0 1 d_y = 1 10 1 .. 15 1 -3 3 15 -3 . -5 -1 -6 5 -5 -6 . d_z = 1 0 1 d_0 = 1 0 1 . 3 1 15 3 1 -3 . 5 -1 -5 5 -1 -6

The columns x, y and z of the original column have been replaced by the column on right hand side of the equation in order to obtain the d_x, d_y and d_z matrix respectively. Calculating the determinants along the top row.

d_x = 10 (+1) 1 -3 + (0)(-1) 15 -3 + (1)(+1) 15 1 . -1 -6 -5 -6 -5 -1 = -90 + 0 + (-10) = -100 d_y = 1(+1) 15 -3 + 10(-1) 3 -3 + 1(+1) 3 15 . -5 -6 5 -6 5 -5 = -105 +30 -90 = -165 d_z = 1(+1) 1 15 + (0) 3 15 + 10(+1) 3 1 . -1 -5 5 -5 5 -1 = 10 + 0 – 80 = -70 d_0 = 1 0 1 = 1 1 -3 - 0 3 -3 + 1 3 1 . . 3 1 -3 -1 -6 5 -6 5 -1 . 5 -1 -6 = -9 – 8 = -17

x= -100/-17 = 5.882 y=-165/-17= 9.706 z = -70/-17 =4.118

Alternatively, the determinants may be calculated in tabular form as shown for d_x thus:

Element	Product of Element and Cofactor	Value
10	10(-6 – 3)	-90
0	- 0(-90 -15)	0
1	1(-15 - - 5)	-10
	Sum	-100

Example 2 Calculate the values of x, y and z by applying Cramer's rule: 10y + x -3z +9 =0 3z +y –x +3 =0 3y + 2x = z -1

Rearranging: x + 10y - 3z = -9 . -x +y + 3z = -3 . 2x + 3y - z = -1

1 10 -3 -9 -1 1 3 = -3 2 3 -1 -1

d_0 = 1 10 -3 ; d_x= -9 10 -3 . -1 1 3 -3 1 3 . 2 3 -1 -1 3 -1

d_y= 1 -9 -3 ; d_z = 1 10 -9 . -1 -3 3 -1 1 -3 . 2 -1 -1 2 3 -1

d_0 = (+1)(1)[1(-1) – 3(3)] + (-1)(10)[-1(-1) -2(3)] + (+1)(-3)[-1(3)

-1(2)] = 55 d_x = (+1)(-9)[1(-1) − 3(3)] + (-1)(10)[-1(-3) -3(-1)] + (+1)(-3)[3(3) −1(-1)] = 54 d_y = (+1)(1)[-3(-1) − (-1)(3)] + (-1)(-9)[-1(-1) − 2(3)] + (+1)(-3)[-1(-1) -2(3)] = -60 d_z = (+1)(1)[1(-1) -3(-3)] + (-1)(10)[-1(-1) -2(-3)] + (+1)(-9)[-1(3)-2(1)]= -17 x= 54/55 = 0.0.9818 ; y=-60/55= -1.0909 and z = -17/55 = -0.3091

Exercise 3.2 Solve for x, y and z by applying Cramer's rule: 1. 2x +6y +z+20=0. . 3x+y − z -1 = 0 . x+5y +6z − 10 =0 2. x-z =10. . 3x +7y + 8z = 12 . 4x+2y+9z =3 3. 4z + 6y –x =5 . x-y +8z = -2 . 10z -3x -7y = -1 4. 2z +3y +7x +9 =3 . . 5x -7y +3z -2= -5 . 2z +3x -2y +4= -4

Chapter 4 Trigonometry

4.1 Introduction

Trigonometry defines the relationship between the sides of a right angle triangle and a specific angle. The basic trigonometrical ratios are:

1. $\text{Sin } A = \dfrac{\text{opposite}}{\text{hypotenuse}}$ 2. $\text{Cos } A = \dfrac{\text{adjacent}}{\text{hypotenuse}}$ 3. $\text{Tan } A = \dfrac{\text{opposite}}{\text{adjacent}}$

Given that in triangle ABC: AB = BC=AC = 2 units AD=DB= 1 unit and CD=$3^{0.5}$ triangle XYZ: XY=YZ=1unit and XZ= $2^{0.5}$ We shall now take a look at the sine, cosine and tan of the special angles in Fig 4.1 and Fig 4.2 as shown in table 4.1. The angles 30, 45 and 60 are known as special angles for two reasons. Their trigonometric ratios can be calculated without using a calculator. Secondly, angles in the second, third and fourth quadrant can be written in terms of the reference angle that these angles make with the positive or negative x axes.

C Z

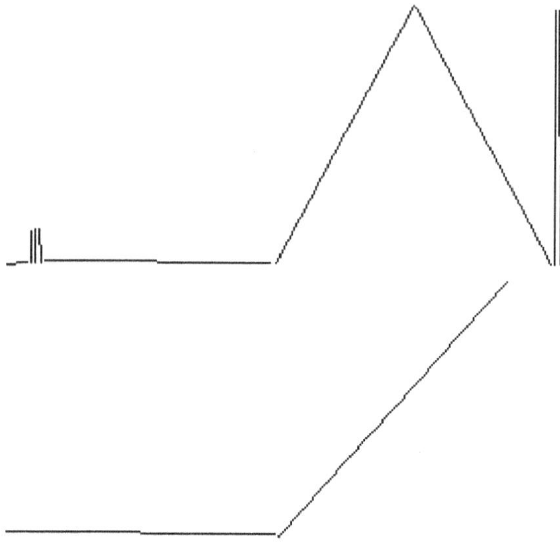

30 30 45

—60 60 45

—A B X Y
Equilateral Triangle Isosceles Triangle
Fig 4.1 Fig 4.2

Table 4.1	30	45	60
Sin	$\dfrac{1}{2}$	$\dfrac{1}{\sqrt{2}} = \dfrac{\sqrt{2}}{2}$	$\dfrac{\sqrt{3}}{2}$
Cos	$\dfrac{\sqrt{3}}{2}$	$\dfrac{1}{\sqrt{2}} = \dfrac{\sqrt{2}}{2}$	$\dfrac{1}{2}$
tan	$\dfrac{1}{\sqrt{3}} = \dfrac{\sqrt{3}}{3}$	1	$\dfrac{\sqrt{3}}{2}$

The reference angle of 120^0 degrees in the second quadrant is 60^0. Likewise the reference angle of 240^0 in the third quadrant is 60^0 because it makes an angle of 60^0 with the negative x axis. The trigonometric ratios sin 120^0, cos 120^0 and tan 120^0 can be written in terms of the special angle 60^0. Fig. 4.3 and table 4.2 below explains this concept.

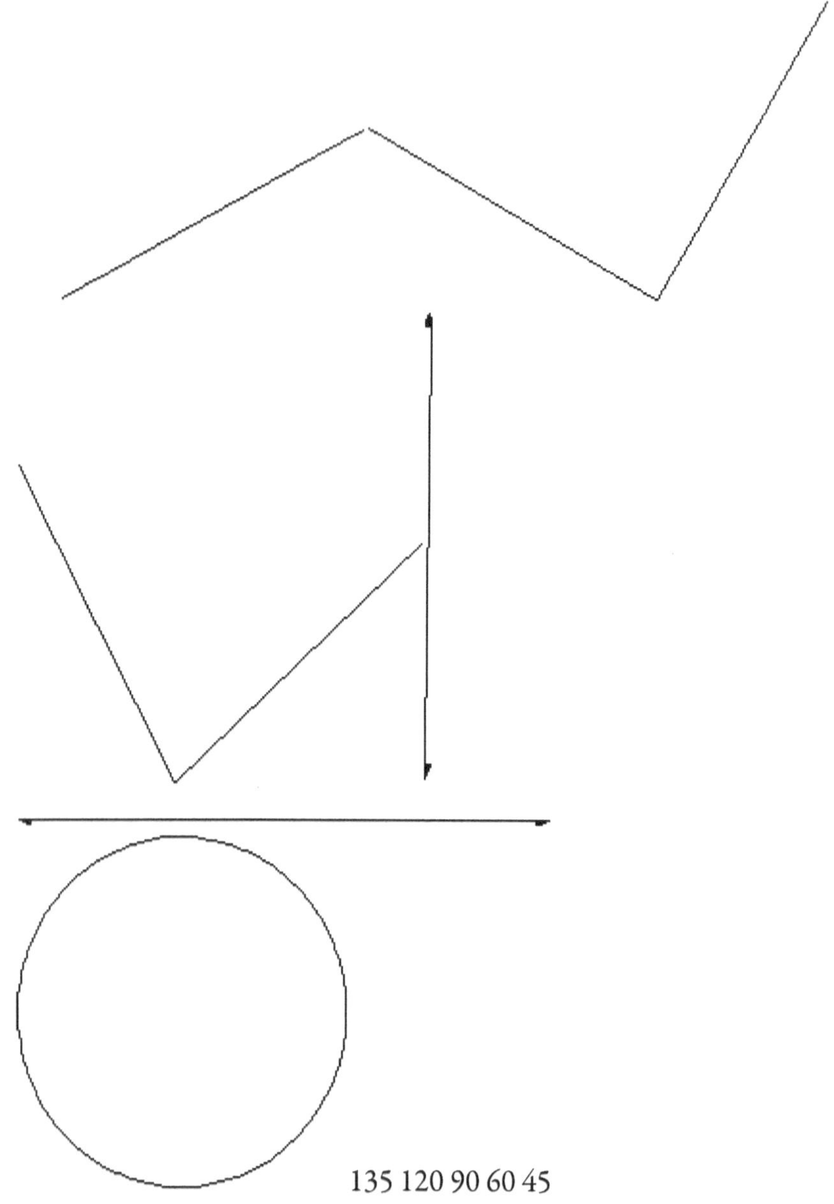
135 120 90 60 45

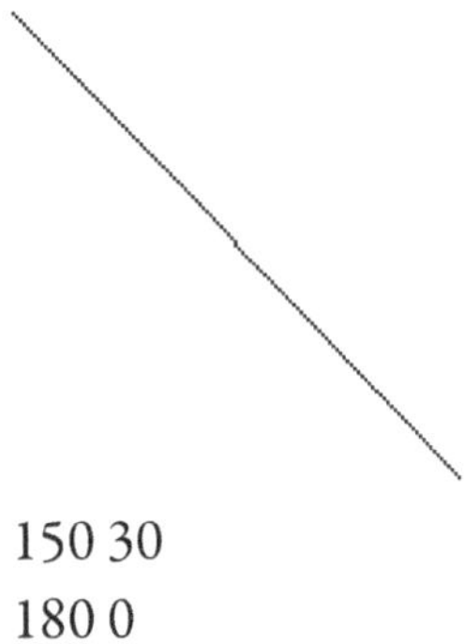

150 30
180 0

210 330
225 240 270 300 315

Fig 4.3 Positive Angles and Reference Angles The table below can be explained thus: sin 120 = sin60 cos 120 = -cos 60 tan 120 = -tan60 As explained above, students are required to write each of the angles in the 2nd, 3rd and 4th columns in terms of their respective reference angle.

4.2 CAST Diagram The CAST Diagram can be clearly defined by: C = Cos : cos of all angles in the fourth quadrant are positive A= All : the cos, sin and tan of all acute angles in the first quadrant are positive S= Sin: the sin of all angles in the second quadrant are positive T= Tan: the tan of all angles in the third quadrant are positive. Study the Fig 4.4 below in order to get a better understanding of the CAST Diagram. The table 4.2 also illustrates the CAST diagram. Table 4.2

Angle	120	135	150	210	225
sin	sin60	sin45	Sin30	-sin30	-sin45
cos	-cos60	-cos45	-cos30	-cos30	-cos45
tan	-tan60	-tan45	-tan30	tan30	tan45

Angle	240	300	315	330	360
sin	-sin60	-sin60	-sin45	-sin30	sin0
cos	-cos60	cos 60	cos45	cos30	cos0
tan	tan60	-tan60	-tan45	-tan30	tan0

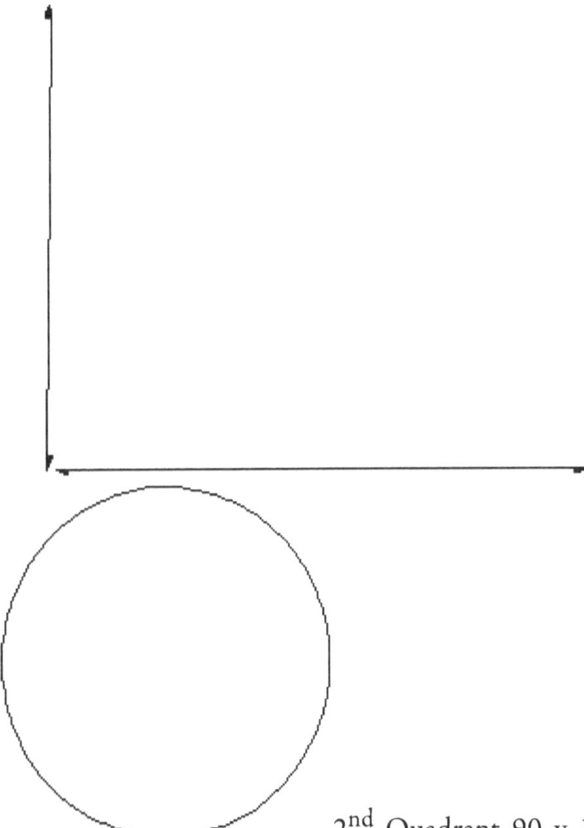

Fig. 4.4 CAST Diagram

Looking at the table 4.2 in the column 330^0 and comparing it with the cast diagram, it is very clear that: cos 330 = 0.866 cos 330 = cos 30 (as stated on the cast diagram) The other trigonometric ratios are negative as shown in the table: tan 330 = -0.5774 tan 330 = -tan 30 sin 330 = -0.5 sin 330 = - sin 30

4.3 Positive and Negative Angles Fig 4.3 and Fig 4.4 will be used to illustrate the measurement of an angle. Positive angles are angles measured from the positive x axis in a counter clockwise direction. Negative angles on the other hand are angles measured in the clockwise direction from the positive x axis. Based on these two definitions of positive and negative angles, a positive angle can be equal to a negative angle. A very good example is the angle 330^0. The angle 330^0 is equal to -30^0.

4.4 Trigonometric Identities A trigonometric identity is an equation that is made up of trigonometric functions. Consider the following trigonometric identities.

1. $\sin^2 A + \cos^2 A = 1$
2. $1 + \cot^2 A = \csc^2 A$
3. $\text{Tan}^2 A + 1 = \sec^2 A$

Substituting the values of A= 30, 45 and 60 respectively into the first equation will give us a value of one. We have substituted the values of the special angles in this case for the sake of simplicity. However, any value substituted into the equation 1 will give us the final answer of 1. For example: $\sin^2 15 + \cos^2 15 = 1$ $\sin^2 300 + \cos^2 300 = 1$

Equation 2 was derived by dividing the LHS and RHS of Equation 1 with $\cos^2 A$. Similarly, dividing Equation 1 by $\sin^2 A$ will give us Equation 3.

The above equations are the most basic trigonometric identities.

4.5 Compound Angles

In the previous sections, we considered the sin, cos and tan of an angle, A. Compound angles is the sin, cos or tan of the sum of two angles. Mathematically, the sine of the sum of two angles is written thus: sin (A+ B)

In addition to the first three identities given above, there are six other compound angle formula as stated below. Sin (A +B) is referred to as a compound angle because there are two angles A and B. Sin (A +B) may be referred to as the sine of the sum of the two angles, A and B.

1. $\mathrm{Sin}(A + B) = \sin A \cos B + \sin B \cos A$
2. $\mathrm{Sin}(A - B) = \sin A \cos B - \sin B \cos A$
3. $\mathrm{Cos}(A + B) = \cos A \cos B - \sin A \sin B$
4. $\mathrm{Cos}(A - B) = \cos A \cos B + \sin A \sin B$
5. $\mathrm{Tan}(A + B) = \dfrac{\tan A + \tan B}{1 - \tan A \tan B}$
6. $\mathrm{Tan}(A - B) = \dfrac{\tan A - \tan B}{1 + \tan A \tan B}$

Example 1. Prove the compound angle formula: $\cos(A - B) = \cos A \cos B + \sin A \sin B$

The distance of the line NM will be calculated by applying the cosine rule and distance formulas. (See Fig 4.5 below) Cosine Rule: $NM^2 = ON^2 + OM^2 - 2(ON)(OM)\cos(A - B)$ Equ. 1 The line NM is opposite the angle $(A - B)$ and the angle $(A - B)$ is bounded by the two sides ON and OM.

Distance Formula: $NM^2 = (\cos A - \cos B)^2 + (\sin A - \sin B)^2$ Equ. 2 The distance formula is applied in order to calculate the length of the line NM. This formula is actually the Pythagoras theorem. $X_N = \cos A$ and $Y_N = \sin A$ $X_M = \cos B$ and $Y_M = \sin B$ Therefore the Point $N = (\cos A, \sin A)$ and Point $M = (\cos B, \sin B)$. The expression $(\cos A - \cos B)$ refers to the base of the triangle PNM while the expression $(\sin A - \sin B)$ is the vertical height of the triangle PNM. Expanding and simplifying the distance formula, we have: $NM^2 = (\cos A - \cos B)^2 + (\sin A - \sin B)^2$

$$\cos^2 A - 2\cos A\cos B + \cos^2 B + \sin^2 A - 2\sin A\sin B + \sin^2 B$$

$$= (\cos^2 A + \sin^2 A) + (\cos^2 B + \sin^2 B) - 2(\cos A\cos B + \sin A\sin B)$$

But $\cos^2 A + \sin^2 A = 1$ and $\cos^2 B + \sin^2 B = 1$ Therefore, $1 + 1 - 2(\cos A\cos B + \sin A\sin B) = 2 - 2(\cos A\cos B + \sin A\sin B)$ From Equ. 1, assuming a unit radius of 1 unit, ON = OM = 1 unit and substituting:

$$NM^2 = 1^2 + 1^2 - 2(1 \times 1)\cos(A - B) \quad NM^2 = 2 - 2\cos(A - B)$$

Equating Equ. 1 to Equ. 2 and simplifying, we have: $2 - 2\cos(A - B) = 2 - 2(\cos A\cos B + \sin A\sin B)$

$$\cos(A - B) = \cos A\cos B + \sin A\sin B$$

Example 2 Evaluate sin 30 Sin 30 = sin (90 − 60) sin90cos60 −

$$\sin60\cos90 = (1 \times 0.5) - \left(\frac{\sqrt{3}}{2} \times 0\right) = 0.5$$

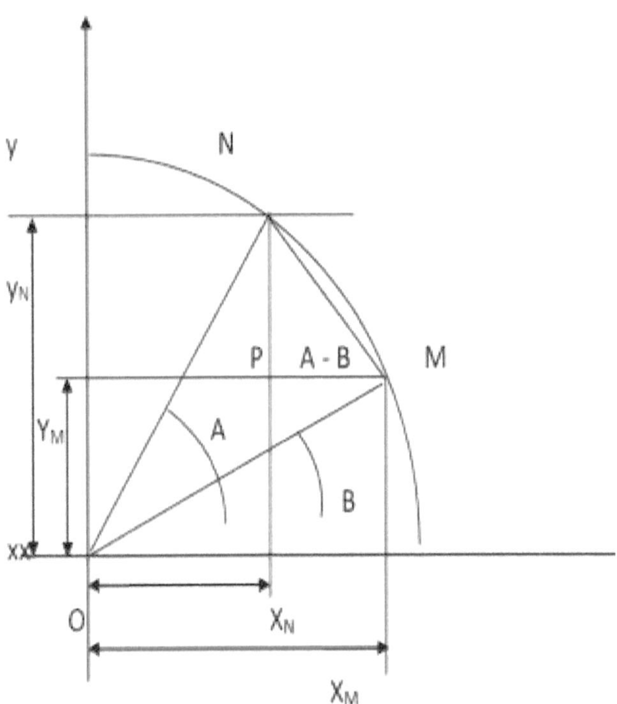

Fig 4.5 Compound Angles

Example 3 Evaluate Cos 75 without using a calculator Cos 75 = Cos

$(45 + 30) = \text{Cos}45\ \text{Cos}30 - \text{Sin}45\text{Sin}30 = (\frac{\sqrt{2}}{2} \times \frac{\sqrt{3}}{2}) - (\frac{\sqrt{2}}{2} \times \frac{1}{2})$

$= \frac{\sqrt{2}}{2}(\frac{\sqrt{3}}{2} - \frac{1}{2}) = 0.2588$

Example 4 Calculate the value of Tan 135.

$\text{Tan } 135 = \text{Tan }(90 + 45) = \frac{\text{Tan }90 + \text{Tan}45}{1 - \text{Tan}90\text{Tan}45} = \frac{\infty + 1}{1 - (\infty \times 1)} = \frac{\infty}{-\infty} = -1$

4.6 Reduction Formulas

Reduction Formulas as the name implies, the sine, cosine or tan of an angle in the second, third or fourth quadrant can be written in terms of any of the special angles 30^0, 45^0 or 60^0.

Sin 30 = 0.5 Sin (90 + 30) = Sin 120 = Sin 90 Cos 30 + Cos 90

$$\text{Sin } 30 = (1 \times \tfrac{\sqrt{3}}{2}) + \left(0 \times \tfrac{1}{2}\right) = \tfrac{\sqrt{3}}{2}$$

Therefore, Sin (90 + 30) =Sin 120 = cos 30 Generally, Sin (90 + Θ) = cos Θ (I) Likewise, Sin (90 – 30) = Sin 60 = Sin 90 Cos 30 – Cos

$$90 \text{ Sin}30 = (1 \times \tfrac{\sqrt{3}}{2}) - (0 \times 0.5) = \tfrac{\sqrt{3}}{2}$$

Therefore, Sin (90 –Θ) = cos Θ (II)

Cos (90 + Θ) = -Sin Θ (III)

Cos (90 – Θ) = Sin Θ (IV)

Tan (90 + Θ) = - Cot Θ (V)

Tan (90 – Θ) = Cot Θ (VI)

Sin (180 + 30) = Sin 210 = Sin180 Cos 30 + Sin 30 Cos 180 = (0

$$X \tfrac{\sqrt{3}}{2}) + (0.5 \times -1) = -\tfrac{1}{2}$$

The equations II, III, IV, V and VI may be evaluated or verified by applying or substituting into the compound angle formula.

4.7 Half Angle Identity The half angle formula shows the relationship between the angle A and $\frac{A}{2}$. $\frac{A}{2}$ is known as the half angle since $A = \frac{A}{2} + \frac{A}{2}$. This formula is derived from the equation $\cos(A + A) = \cos^2 A - \sin^2 A$, where $A = \frac{A}{2}$. Substituting $A = \frac{A}{2}$ $\cos(\frac{A}{2} + \frac{A}{2}) = \cos^2\frac{A}{2} - \sin^2\frac{A}{2}$ and substituting $\sin^2\frac{A}{2} = 1 - \cos^2\frac{A}{2}$, we have: $\cos^2\frac{A}{2} - \sin^2\frac{A}{2} = \cos^2\frac{A}{2} - (1 - \cos^2\frac{A}{2}) = 2\cos^2\frac{A}{2} - 1$ Therefore $\cos A = 2\cos^2\frac{A}{2} -$

1 $2\cos^2\frac{A}{2} = 1 + \cos A$ $\cos^2\frac{A}{2} = \frac{1}{2}[1 + \cos A]$ $\cos\frac{A}{2} = [\frac{1}{2}[1 + \cos A]]^{0.5}$

Similarly, $\sin\frac{A}{2} = [\frac{1}{2}[1 - \cos A]]^{0.5}$ may be derived.

Example 1 Evaluate $\sin 75^0$ and $\cos 75^0$.

$\sin(75 + 75) = \sin 150$ $\sin 75 = \sin(2 \times 75) = \sin 150$ $A = 150^0$

and $\frac{A}{2} = 75^0$ Corresponding reference angle $= 180 - 150 = 30^0$ Cos

$A = 2\cos^2\frac{A}{2} - 1$ $\cos\frac{A}{2} = [\frac{1}{2}(1 + \cos A)]^{0.5} = [\frac{1}{2}(\cos 150 +$

$1)]^{0.5} = 0.2588$ $\sin\frac{A}{2} = [\frac{1}{2}(1 - \cos A)]^{0.5} = [\frac{1}{2}(1 - \cos 150)]^{0.5}$

$= 0.9659$

4.8 Double Angle Identity

$\sin(A + B) = \sin A\cos B + \sin B\cos A$ given that $A = B$
$\sin(A + A) = \sin 2A$ $\sin(A + A) = \sin A \sin A + \sin A\cos A$
$= 2\sin A\cos A$

Similarly, $\cos(A + B) = \cos A\cos B - \sin A\sin B$

$\cos(A + A) = \cos 2A$ $\cos(A + A) = \cos A\cos A - \sin A\sin A$

$= \cos^2 A - \sin^2 A$

$= \cos^2 A - (1 - \cos^2 A) = 2\cos^2 A - 1$

$= (1 - \sin^2 A) - \sin^2 A = 1 - 2\sin^2 A$

$\tan 2A = \dfrac{2\tan A}{1 - \tan^2 A}$

Example 1. Evaluate $\cos 90$ Given that $A = 45^0$ and $\cos A =$

$2^{-0.5}$

$$\cos 90 = \cos (2 \times 45) = 2\cos^2 45 - 1 = 2 \times \left[\frac{1}{\sqrt{2}}\right]^2 - 1 = 0$$

Example 2. Evaluate tan 2A and cos 2A given that $\tan A = \sqrt{3}$ and $\sin A = \frac{\sqrt{3}}{2}$. Leave your answer in surd form. Sketch the right-angled triangle and solve for the unknown side as shown below.

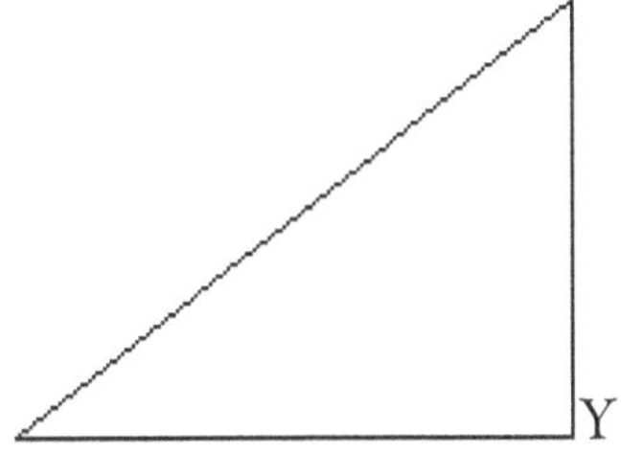

$$2\ 3^{0.5}$$

$$X\ A\ Z$$

Fig. 4.6

$$XZ = (3 - 2)^{0.5} = 1 \text{ unit}$$

$$\text{Tan } 2A = \frac{2\text{Tan } A}{1-(\tan^2 A)} = \frac{2 \times \sqrt{3}}{1-(\sqrt{3})^2} = -\sqrt{3}$$

$$\text{Cos } 2A = \frac{\sin\sin 2A}{\tan 2A} \quad \sin 2A = 2\sin A \cos A \quad 2\sin A\cos A = 2 \times \frac{\sqrt{3}}{2} \times \frac{1}{2}$$

$$= \frac{\sqrt{3}}{2} \quad \text{Cos } 2A = \frac{\frac{\sqrt{3}}{2}}{-\sqrt{3}} = -\frac{1}{2}$$

Exercise 4.1

1. Evaluate Sin2A given that 0<A<90

    i. $\text{Sin } A = \frac{7}{12}$ ii) $\cos A = \frac{3}{8}$ iii) $\tan A = \frac{3}{8}$

1. Given the formula $\cos 2A = 2 \cos^2 A - 1$, evaluate the following by firstly calculating A:

    i. cos 15 ii) cos 37.5 iii) cos 22.5

1. Evaluate the following by applying the corresponding double angle identity:

    i. Sin 60 ii) sin 120 iii) tan 90

1. Calculate the value of Sin 2A, Cos 2A, Tan 2A and Cot 2A given that sin A = 0.9659.
2. Calculate the value of the following given that sin A = 0.5 (0<A<360)

   i. Cosec 2A ii) sec 2A iii) cot 2A

1. Calculate the value of the following given that sin A = - 0.5 (0<A<360) i)Cosec 2A ii) sec 2A iii) cot 2A
2. Evaluate Cos 2A and tan 2A given that (0<A<360) i) Sin A $= \frac{7}{12}$ ii)cosA $= \frac{3}{8}$ iii)tanA $= \frac{3}{8}$
3. Evaluate cos 292.5 and tan 292.5 without using a calculator.
4. Evaluate sin 337.5 and tan 337.5 without using a calculator.
5. Evaluate cos 345 and tan 345 without using a calculator.
6. Calculate tan A if sin [A/2] = $2/6^{0.5}$ and cos [A/2] = $1/3^{0.5}$
7. Find cot A given that sin [A/2] = 1/4.

4.9 Trigonometric Equations

Trigonometric Equations are solved by considering the positive and negative signs of a trigonometric ratio in the CAST diagram. The positive or negative sign determines the quadrant the angle lies in. Trigonometric ratios may be categorized thus:

a) y = sin Θ Simplest form b) y = $9\sin^2 Θ$ – 6sin Θ + 1 Quadratic form c) i) 6sin2Θ = cos 2Θ Double Angle ii) $\tan^2 2Θ = 2\sin2Θ$ iii) 2cos Θ + sin2Θ = 0 d) Simultaneous Trigonometric Equations i) y = tan $\frac{x}{2}$ y = $2\sin\frac{x}{2}$

ii) y = $\sin\frac{x}{4}$ y = $3\sin\frac{x}{2}$ In b) above, we are given a quadratic formula in which sin Θ is the unknown. The Quadratic Formula is applied to solve this equation by substitution of the values of a, b and c as shown in Example 4.

It is common to find equations such as $y = 9\cos^2 \Theta - 6\sin \Theta + 1$. In this case, there are two unknowns namely $\cos \Theta$ and $\sin \Theta$. The relationship between the two unknowns is: $\sin^2\Theta + \cos^2 \Theta = 1$.

Substitution of $\cos^2 \Theta = 1 - \sin^2 \Theta$ will ensure that we have a trigonometric quadratic formula having only one unknown ($\sin \Theta$).

The third category of trigonometric equations is one in which we have to apply our knowledge of compound angles, double angle and half angle formula in order to solve the problem.

The two pairs of equations in the fourth category are referred to as simultaneous trigonometric equations.

Example 1 Calculate all possible values of $\cos \Theta = \sec \Theta$ between 0^0 and 360^0.

$\cos \Theta = \dfrac{1}{\cos\cos \Theta}$ multiplying both sides by $\cos \Theta$, we have $\cos^2 \Theta = 1$ taking square root of both sides $\cos \Theta = \pm 1$ Referring to the CAST diagram, cos is positive in the 4th Quadrant and negative in the 2nd and 3rd Quadrant. $\Theta = 0^0$; $\Theta = 180$ The positive and negative x axis which is also the equivalent of the $0^0 - 180^0$ line, differentiates the 1st and 4th quadrants and the 2nd and 3rd quadrants respectively.

Example 2 Evaluate all possible values of $6\sin2\Theta = \cos2\Theta$

Simplifying, $\dfrac{\sin\sin 2\Theta}{\cos 2\Theta} = \dfrac{1}{6}$ thus $\tan2\Theta = \dfrac{1}{6}$ 1st Quadrant $2\Theta = 9.4623^0$ $\Theta = 4.7312^0$ 3rd Quadrant $2\Theta = 180^0 + 9.4623^0 = 189.4623^0$ $\Theta = 94.7312^0$ Tan is +ve in the 1st and 3rd quadrants. Students should observe carefully the expression $2\Theta = 180^0 + 9.4623^0$ as against the expression $\Theta = 180^0 + 9.4623^0$. $\Theta = 180^0 + 9.4623^0$ is the angle in the 3rd quadrant when our unknown angle is Θ. Alternatively, if we are given an original equation $6\sin\Theta = \cos\Theta$, our angle in the 3rd quadrant would be: $\Theta = 180^0 + 9.4623^0$.

Example 3 Evaluate all values of Θ for $0^0 \leq \Theta \leq 360^0$ given that $2\cos\Theta + \sin2\Theta = 0$

$2\cos\Theta + \sin\Theta\cos\Theta = 0$ $\cos\Theta(1 + \sin\Theta) = 0$ $\cos\Theta = 0, \Theta = 90^0$ $1 + \sin\Theta = 0$ $\sin\Theta = -1$. Therefore, $\Theta = 270$ or -90 (negative angle)

Example 4 Given that $9\sin^2\Theta - 6\sin\Theta + 1 = 0$, calculate Θ for $0^0 \leq \Theta \leq 360^0$.

$9\sin^2\Theta - 6\sin\Theta + 1 = (3\sin\Theta - 1)(3\sin\Theta - 1)$ $(3\sin\Theta - 1)$

$= 0$ $\Theta = \sin^{-1}(\frac{1}{3}) = 19.4712^0$ (1st Quadrant) $\Theta = 180^0 - 19.4712^0$ $= 160.5288^0$ (2nd Quadrant) Sine is positive in our 1st and 2nd quadrants. Alternatively, this problem may be solved by substituting the values of $a = 9$, $b = -6$ and $c = 1$ into the quadratic formula.

.**Example 5** $\tan^2 2\Theta = 2\sin 2\Theta$ calculate Θ for $0^0 \leq \Theta \leq 360^0$

$\sin^2 2\Theta = (2\sin2\Theta).(\cos^2 2\Theta)$ $\sin2\Theta = 2(\cos^2 2\Theta)$ since $\tan^2 2\Theta$ $= (\sin^2 2\Theta) \div (\cos^2 2\Theta)$ $\cos^2 2\Theta + \sin^2 2\Theta = 1$ $\cos^2 2\Theta = 1 - \sin^2 2\Theta$ Simply and substitute by evaluating based on one unknown $\sin2\Theta = 2(1 - \sin^2 2\Theta)$ $2\sin^2 2\Theta + \sin 2\Theta - 2 = 0$ Quadratic trigonometric equation in terms of 2Θ. $\sin 2\Theta = \dfrac{-1 \pm \sqrt{1^2 - 4(2)(-2)}}{2(2)}$ $= \sin 2\Theta = 0.7808$ $2\Theta = 51.3339^0$ $\Theta = 25.6669^0$ (1st Quadrant) $2\Theta = 180 - 51.3339^0$ $= 128.6661$ $\Theta = 64.3331$ (2nd Quadrant) $\sin 2\Theta = -1.2808$ (not possible) $\sin 2\Theta$ cannot be less than -1 and $-1 \leq y = \sin 2\Theta \leq +1$

4.10 Trigonometric Equations: Rotations and Number of Revolutions The table below summarizes the effect of rotating a given angle through a number of revolutions about the origin. Alternatively, this can be considered as $\sin (A + 180)$, $\sin (A +$

360) or sin (A + 720) as the case maybe. Angle, A, has been taken as 30^0 and 60^0 for table 4.3 below. However, the table applies to any angle that is less than 90^0.

Table 4.3

Value		Compound Angle	Value	Number of Revolutions
Sin 30	0.5	Sin (30 + 180)	-0.5	0.5
Cos 30	0.866	Cos (30 +180)	-0.866	0.5
Tan 30	0.577	Tan (30 + 180)	0.577	0.5
Sin 30	0.5	Sin (30 + 360)	0.5	1
Cos 30	0.866	Cos (30 +360)	0.866	1
Tan 30	0.577	Tan (30 + 360)	0.577	1
Sin 30	0.5	Sin(30 +720)	0.5	2
Cos 30	0.866	Cos(30 + 720)	0.866	2
Tan 30	0.577	Tan(30+720)	0.577	2

The application of the CAST Diagram verifies or confirms the positive or negative value of our answers. For the first three rows in table 2.3, rotation through half a revolution about the center is equivalent to sum of two angles or the rotation about the center (the third quadrant). In the third quadrant, sine and cos are negative while tan is positive.

The solution is based on applying this procedure for rotation of angle 30^0 through one revolution and two revolutions are 390^0 ($30^0 + 360$) and 750^0 ($30^0 + 720^0$).

The calculator evaluations of $\sin 390^0$ and $\sin 750^0$ as stated in the table is a standard exercise for students. Practice this evaluation for $\cos 390^0$ and $\cos 750^0$.

In conclusion, the sin, cos and tan of the angles $A+360^0$, $A + 720^0$ and (A + k revolutions) are the sames value of sin A, cos A and tan A. This is true given that k = 1, 2, 3, 4, 5, 6, 7...k complete revolutions.

4.11. General Formula for Trigonometric Equations

Based on the conclusion of the previous section, the general formula of Examples 1 – 5 may be summarized in the table below.

Table 4.4

	Problem	Value of	General Formula
Ex. 1	$Cos^2\Theta = 1$	$\Theta = 0$	$\Theta = 0 + k$
		$\Theta = 180$	$\Theta = 180 + k$
Ex. 2	$6sin2\Theta = cos2\Theta$	$2\Theta = 9.4623$	$\Theta = 4.7312 + k$
		$2\Theta = 180 + 9.4623$	$\Theta = 90 + 4.7312 + k$
Ex. 3	$2cos\,\Theta + sin\,\Theta cos\,\Theta = 0$	$\Theta = 270$	$\Theta = 270 + k$
		$\Theta = 0$	$\Theta = 0$
Ex. 4	$9sin^2\Theta - 6sin\,\Theta + 1 = 0$	$\Theta = 19.4712$	$\Theta = 19.4712 + k$
		$\Theta = 180 - 19.4712$	$\Theta = 160.5288 + k$
Ex. 5	$Tan^2 2\Theta = = 2sin2\,\Theta$	$2\Theta = 51.3339$	$\Theta = 25.6669 + k$
		$2\Theta = 180 - 51.3339$	$\Theta = 90 - 25.6669 + k$

K = 1, 2, 3, 4, 5, 6, 7.............k complete revolutions. Substituting these number of revolutions into the original equation will give us the same answer for n=both right hand side and left hand side. For Example 2: $6sin(2\Theta + 360) = cos(2\Theta + 360)$ $6sin(2\Theta + 720) = cos(2\Theta + 720)$

4.12 Simultaneous Trigonometric Equations These trigonometric equations are a given pair of equations which will give the y coordinate when the x values are substituted.

Alternatively, for example 1, substituting both values of x and y into both equations will give the value of 1.

Example 1 Given that sin $(3x + y) = 1$ and tan $(x + y) = 1$, solve for x and y.

$(3x + y) = \sin^{-1}(1) = 90^0$ $\sin^{-1}(1) = 90^0$ $(3x + y) = 90^0$ $(x + y) = \tan^{-1}(1) = 45^0$ $\tan^{-1}(1) = 45^0$ $(x + y) = 45^0$ Therefore, $3x + y = 90$ and $x + y = 45$ $x = 45 - y$ Substituting, $3(45-y) + y = 90$ $y = 22.5^0$ and $x = 45 - y = 22.5$

Example 2 Calculate the values of x and y that satisfies the equations: tan $(x + y) = 3^{0.5}$ and Tan $(3x - y) = 1$.

$(x + y) = \tan^{-1} 3^{0.5} = 60^0$ $x = 60 - y$ $3x - y = \tan^{-1} 1 = 45^0$ $3(60 - y) - y = 45^0$ $y = 33.75^0$ and $x = 60 - 33.75^0 = 26.25^0$

Example 3 Solve for x and y given that: $y = \tan \frac{x}{2}$ and $y = 2\sin\frac{x}{2}$ tan $\frac{x}{2} = 2\sin\frac{x}{2}$ Considering the fact that: $\tan \frac{x}{2} = [\sin \frac{x}{2}] \div [\cos\frac{x}{2}]$ and substituting into the above equation, we have $[\sin\frac{x}{2}] \div [\cos\frac{x}{2}] = 2\sin\frac{x}{2}$ Simplifying: $\cos\frac{x}{2} = 0.5$ 1st Quadrant: $\frac{x}{2} = 60 + k360$ and $x = 120 + k720$ 4th Quadrant: $\frac{x}{2} = 360 - 60 + k360$ and $x = 600 + k720$

Example 4 Solve the simultaneous trigonometric equation: $y = \sin\frac{x}{4}$ and $y = 3\sin\frac{x}{2}$

In this problem, there are two unknowns. The formula below states the relationship between them. $\frac{x}{2} = \frac{x}{4} + \frac{x}{4}$ $y = 3\sin\frac{x}{2}$ can thus be written as follows: $3\sin[\frac{x}{4} + \frac{x}{4}]$ We can now apply the compound angle formula. $3\sin[\frac{x}{4} + \frac{x}{4}] = 3[2\sin\frac{x}{4}\cos\frac{x}{4}]$ From the original equations: $\sin\frac{x}{4} = 3\sin\frac{x}{2}$

Substituting our new expression for $3\sin\frac{x}{2}$ as explained above.

$$\text{Sin}\frac{x}{4} = 6\sin\frac{x}{4}\cos\frac{x}{4}$$

We now have one unknown which is $\frac{x}{4}$.

Simplifying, we have, $\cos\frac{x}{4} = \frac{1}{6}$

1st Quadrant: $\frac{x}{4} = 80.406 + k360$ $x = 321.624 + k360$ 4th

Quadrant: $\frac{x}{4} = 360 - 80.406 + k360$ $x = 1118.376 + k360$

4.13 Solving Trigonometric Identities There are two categories of trigonometric identity problems. In the first category, students are required to prove an equation. The second category is a problem in which learners are required to simplify. These problems are usually written in an exam to test the student's basic understanding of the identity equations in sections 4.4, 4.5, 4.7 and 4.8. **Example 1** Prove that $\dfrac{2\cos\cos\theta}{1+\tan^2\theta} = 2\cos^3\theta$

Simplify the left hand side $\dfrac{2\cos\cos\theta}{1+\tan^2\theta} = \dfrac{2\cos\cos\theta}{\sec^2\theta} = \dfrac{2\cos\cos\theta}{1/\cos^2\theta} = 2\cos^3\theta$

Example 2 Provethat $\sin2^\theta.\cot^\theta.\mathrm{Cosec}^2\theta = \cot^\theta$.

$$\frac{\sin^2\theta.\cos\cos\theta.\,(1)}{\sin\theta.\;\sin^2\theta} = \cot^\theta$$

Example 3. Simplify $(1 - \sin A)(1 + \sin A)$

$1^2 - \sin^2 A = 1 - \sin^2 A = \cos^2 A$

Example 4. Simplify $\dfrac{2+2\cos\cos\theta}{\sin^2\theta}$

$$\frac{2+2\cos\cos\theta}{\sin^2\theta} = \frac{2(1+\cos\cos\theta)}{1^2-\cos^2\theta} =$$

$$\frac{2(1+\cos\cos\theta)}{(1-\cos\cos\theta)(1+\cos\cos\theta)} = \frac{2}{(1-\cos\cos\theta)}$$

Exercise 4.2

1. Solve for the values of θ for $0^0 \le \theta \le 360^0$ given that $5\cos\theta - 7\sin\theta = 0$.

2. Evaluate θ for $0^0 \le \theta \le 360^0$ given that $7\cos\theta - 10\sin\theta = 0$

3. Given that $9\sin^2\theta + 6\sin\theta + 1 = 0$, calculate θ for $0^0 \le \theta \le 360^0$.

4. Solve for the values of θ for $0^0 \le \theta \le 360^0$ given that $5\cos\theta + \sin2\theta = 0$.

5. Solve the trigonometric equation $5\sin^2\theta + 10\sin\theta + 1 = 0$ for $0^0 \le \theta \le 360^0$.

6. Given that $25\tan^2\theta - 1 = 0$, calculate θ for $0^0 \le \theta \le 720^0$.

7. Solve the trigonometric equation $\tan\theta - 5\sec^2\theta + 10 = 0$ for $0^0 \le \theta \le 360^0$.

8. Given that $17 - 10\sec^2\theta + \tan\theta = 0$, calculate θ for $0^0 \le \theta \le 180^0$.

9. Find the values of θ $0^0 \le \theta \le 360^0$ for the equation $4\sin\theta = \csc\theta$.

10. Solve the trigonometric equation $10\cos^2\theta = 1 - 7\sin\theta$ for $0^0 \le \theta \le 360^0$.

11. Given that $36\tan^2\theta - 4 = 0$, calculate θ for $0^0 \le \theta \le 720^0$.

12. Solve for the values of θ for two revolutions $0^0 \le \theta \le 720^0$ given that $\tan\frac{\theta}{2} = \cot\frac{\theta}{2}$

13. Find the values of x and y given that $\sin(2x - y) = 0.5$ and $\cos(x + y) = 1$

14. Solve the simultaneous trigonometric equation $\tan(x + y) = 1$ and $\cot(2x + 3y) = 0$.

15. Solve for x and y given that $\cos(3x - y) = 0$ and $\cos(x + y) = -1$.

16. Find the values of x and y given that $\sin(3x + y) = -0.5$ and $\cos(x + y) = 1$.

17. Prove that $\dfrac{1-\sin\sin A}{\sin\sin A} + \dfrac{2\cos\cos A}{1-\cos\cos A} = \dfrac{1}{\sin} + \dfrac{1-3\cos\cos A}{1-\cos\cos A}$.

18. Simplify $\dfrac{2\tan\tan\theta}{1+\cot^2\theta}$.

19. Simplify $\dfrac{1+2\theta}{\sec^2\theta}$

Chapter 5 Complex Numbers

5.1 Introduction Complex numbers are written in two standard forms.

a. ⎯ Rectangular form $z = x + iy$ where x = real part, iy= imaginary part and $i = (-1)^{0.5}$

b. ⎣⎯ Polar form $z = r(\cos\Theta + i\sin\Theta) = r\Theta$

r = modulus = $(x^2 + y^2)^{0.5}$

Θ = argument $\tan\Theta = y/x$ and $\Theta = \tan^{-1}(y/x)$

5.2 Argand Diagram

This is a graphical representation of a complex number. The vertical axis is the imaginary axis and it known as the iy axis. The horizontal axis is the x-axis. The next section explains the relationship between both forms of complex numbers as stated in section 5.1.

5.3. Rectangular and Polar Complex Numbers. The

Argand diagram Fig.5. 1 will be used as reference. OB = 3, OA = 3, BD = AC = AE = BF = 6

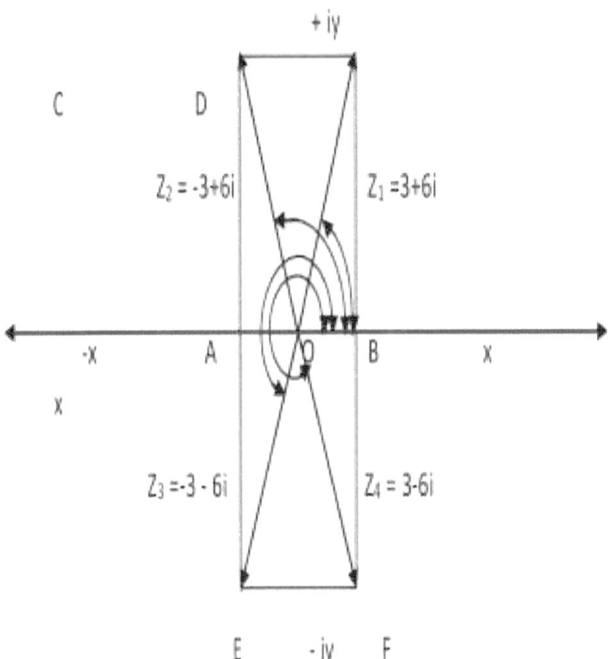

Fig 5.1 Argand Diagram

First Quadrant: Consider complex number $Z_1 = 3 + 6i$, $r = (3^2 + 6^2)^{0.5} = 6.708$ $\Theta = \tan^{-1}(6/3) = 63.435$ $Z_1 = 6.708(\cos 63.435 + i\sin 63.435)$

Second Quadrant: $Z_2 = -3 + 6i$, $r = (3^2 + 6^2)^{0.5} = 6.708$ (ignore the signs) $\Theta_{r2} = \tan^{-1} (6/3)$ $\Theta_{r2} = 63.435$(ignore the signs) $\Theta_{r2} =$ reference angle of Z_2 $\Theta_{r2} = \Theta_{r3} = \Theta_{r4}$ Reference angle is the acute angle between +x or –x axis and the complex number $\Theta_2 = 180 - \Theta_{r2} = 180 - 63.435 = 116.565$ $Z_2 = 6.708$ (cos 116.565+ isin 116.565) Third Quadrant: $Z_3 = -3 - 6i = 6.708$ (cos 243.435+ isin 243.435) $\Theta_3 = 180 + \Theta_{r3} = 180 + 63.435 = 243.435$ Fourth Quadrant: $Z_4 = 3 - 6i = 6.708$ (cos 296 .545+ isin 296 .545) $\Theta_3 = 360 - \Theta_{r4} = 360 - 63.435 = 296 .545$ The table below summarizes the relationship between complex number rectangular form and complex number polar form. The Polar Form column has been written in a general form based on section 5.1 above. Negative angles are measured in a clockwise direction from the positive x axis to the complex number in question.

Table 5.1

Quadrant	Rectangular Form	Polar Form
1st	$Z_1 = 3+6i$	$r_1[\cos \Theta_1 + \sin \Theta_1]$
2nd	$Z_2 = -3+6i$	$r_2[\cos (180-\Theta_{2ref})+ \sin (180-\Theta_{2ref})]$
3rd	$Z_1 = -3-6i$	$r_1[\cos (180+\Theta_{3ref})+ \sin (180+\Theta_{3ref})]$
4th	$Z_1 = 3-6i$	$r_1[\cos (360-\Theta_{4ref})+ \sin (360-\Theta_{4ref})]$

Table 5.2

Rectangular Form	Positive Angles	Negative Angles
$Z_1 = 3+6i$	$\angle 6.708 \; 63.435$	$\angle 6.708 \; -296.565$
$Z_2 = -3+6i$	$\angle 6.708 \; 116.565$	$\angle 6.708 \; -243.435$
$Z_1 = -3-6i$	$\angle 6.708 \; 243.435$	$\angle 6.708 \; -116.565$
$Z_1 = 3-6i$	$\angle 6.708 \; 296.545$	$-63.435 \; \angle 6.708$

Table 5.2 shows the equivalent negative angle complex number of each polar complex number as evaluated earlier.

5.4 Quadratic Complex Number Given the quadratic formula, $y = \dfrac{-b \pm \sqrt{b^2-4ac}}{2a}$, the term $(b^2 - 4ac)$ gives us an undefined or error value when b^2 is less than $4ac$. This is due to the fact that the square root of a negative number is undefined or error. However, the expression can be written in terms of -1 or square root of-1 as stated below. $(b^2 - 4ac) = (-1(-b^2 +4ac)) = (-1 \times (4ac - b^2)^{0.5} = (-1)^{0.5} \times (4ac - b^2)^{0.5} = i(4ac - b^2)^{0.5}$ The original quadratic formula reduces to a complex number (real and imaginary components). $y = \dfrac{-b \pm i\sqrt{4ac-b^2}}{2a}$ where $\dfrac{-b}{2a}$ = real component and $\dfrac{i\sqrt{4ac-b^2}}{2a}$ = imaginary component

_______________**5.5 Addition and Subtraction of Complex numbers** **Example 1:** Given the complex numbers: $Z_1 = 3 + 10i$ $Z_2 = -4 +7i$ $Z_3 = -9 -i$ $Z_4 = 3 -6i$ evaluate the following:

a. $Z_1 + Z_2$

b. $Z_3 - Z_4$

c. $Z_1 + Z_2 + Z_3 - Z_4$

Solution:

a. $Z_1 + Z_2 = (3 + 10i) + (-4 + 7i) = (3 - 4) + (10i + 7i) = -1 + 17i$

b. $Z_3 - Z_4 = (-9 - i) - (3 - 6i) = -9 - 3 - i + 6i = -12 + 5i$

c. $Z_1 + Z_2 + Z_3 - Z_4 = (3 + 10i) + (-4 + 7i) + (-9 - i) - (3 - 6i)$

$$= 3 - 4 - 9 - 3 + 10i + 7i - i + 6i = -13 + 22i$$

5.6 Multiplication and Division of Complex Numbers: Rectangular Form

Example 2: Given the complex numbers $Z_1 = 3 + 10i$ $Z_2 = -4 + 7i$ $Z_3 = -9 - I$ $Z_4 = 3 - 6i$ evaluate the following:

a. $Z_1 \times Z_2$

b. $Z_1 \times Z_2 \times Z_4$

c. $Z_1 \div Z_2$

Solution:

a. $Z_1 \times Z_2 = (3 + 10i) \times (-4 + 7i) = 3(-4 + 7i) + 10i(-4 + 7i)$

$$= -12 + 21i - 40i + 70i^2 = -12 - 19i + 70(-1) = -82 - 19i$$

a. $Z_1 \times Z_2 \times Z_4 = (-82 - 19i)(3 - 6i) = -82(3 - 6i) - 19i(3 - 6i)$

$$= -246 + 492i - 57i + 114i^2 = -246 + 435i + 114(-1) = -360 + 435i$$

a. $Z_1 \div Z_2 = \dfrac{(3 + 10i)(-4-7i)}{(-4 +7i)(-4-7i)}$

$= \dfrac{3(-4-7i)+ 10i(-4-7i)}{(-4)(-4))-(-7i)(-7i)}$

$= \dfrac{-12-21i-40i-70(-1)}{16-49(-1)}$

$= \dfrac{58-61i}{65}$

$= 0.8923 - 0.9385i$

-4 -7i is known as the conjugate of -4 + 7i.

The product of a complex number and its conjugate is the square of the modulus. $(-4 -7i) \times (-4 + 7i) = 4^2 - i^2\, 7^2 = r^2\, 4^2 - (-1)^2\, 7^2 = r^2$ The importance of multiplying the numerator and the denominator of a complex number in fraction form is that it eliminates the imaginary component in the denominator. Refer example 3 above.

                     EFETOBO EMEDE

5.7 Multiplication and Division of Polar Complex Numbers

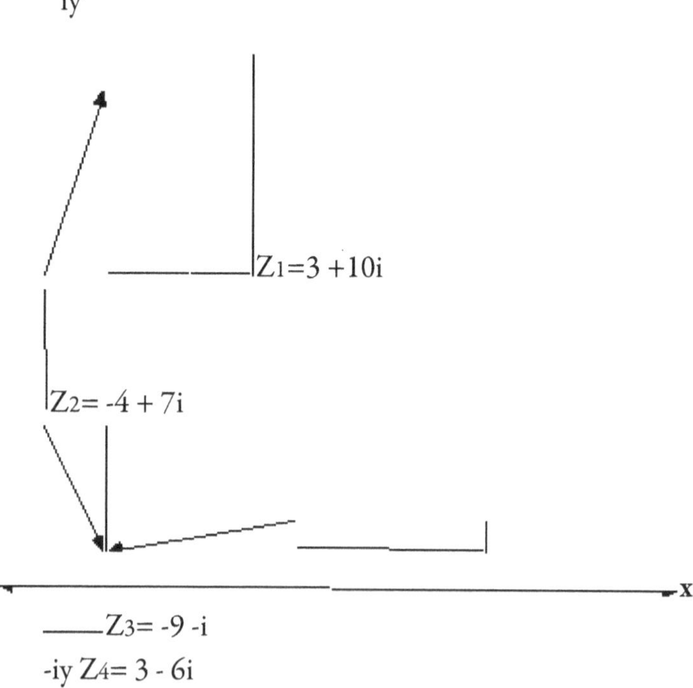

Fig 5.2 Argand Diagram

Rectangular complex numbers as given: $Z_1 = 3 + 10i$ $Z_2 = -4 +7i$ $Z_3 = -9 -i$ $Z_4 = 3 -6i$ The conversion of the rectangular complex number into their corresponding polar complex numbers is based on the argand diagram as illustrated. Refer to Fig 5.2.

Arguments are measured as positive angles anti -clockwise from the positive x axis to the corresponding complex number.

$\lfloor\underline{\hspace{2cm}}$ $Z_1 = 3 + 10i$: $r_1 = (3^2 + 10^2)^{0.5} = 10.44$ and $\Theta_1 = \tan^{-1}$ $(10/3) = 73.3^0$ $Z_1 = 10.44\ 73.3$

$Z_2 = -4 +7i$: $r_2 = (4^2 + 7^2)^{0.5} = 8.062$ $\Theta_{r2} = \tan^{-1} (7/4) = 60.255$ $\Theta_2 = 180 - 60.255 = 119.745$

$\underline{\hspace{1.5cm}}\lfloor$ $Z_2 = 8.062\ 119.745$

$Z_3 = -9 -i$: $r_3 = (9^2 + 1^2)^{0.5} = 9.055$, $\Theta_{r3} = \tan^{-1} (1/9) = 6.34$ $\Theta_3 = 180 + 6.34 = 186.34$

$\lfloor\underline{\hspace{2cm}}\rfloor$ $Z_3 = 9.055\ 186.34$

$\underline{\hspace{2.5cm}}$ $Z_4 = 6.708\ 296.545$

Example 3

1. $Z_1 \times Z_2 = 10.44\ 73.3 \times 8.062\ 119.745$

$\underline{\hspace{5cm}}\|\underline{\hspace{1cm}} = 10.44 \times 8.062\ 73.3 + 119.745 = 84.167\ 193.045$

$84.167 (\cos 193.045 + i\sin 193.045) = -81.995 -18.998i$

$\lfloor\underline{\hspace{5cm}}\rfloor$ 2. $Z_1 \div Z_2 = \dfrac{10.44\ \text{cis}\ 73.3}{8.062\ \text{cis}\ 119.745} =$

$\dfrac{10.44}{8.062}\ 73.3 - 119.745 = 1.295\ -46.445$

$295 (\cos (-46.445) + i\sin(-46.445)) = 0.8923 - 0.9385i$

$\underline{\hspace{2cm}}\lfloor\underline{\hspace{1.5cm}}\lfloor\underline{\hspace{2cm}}\rfloor$ 3. $Z_1 \times Z_2 \times Z_4 = 10.44\ 73.3 \times 8.062\ 119.745 \times 6.708\ 296.545$

$\underline{\qquad\qquad\qquad\qquad\qquad|\qquad|} = 10.44 \text{ X } 8.062 \text{ X } 6.708$

$73.3 + 119.745 + 296.545 = 564.5941\ 489.59 = 564.5941(\cos 489.59 + i\sin 489.59) = -359.8099 + 435.09i$

General Formula for Multiplication and Division (Polar Form)

1. $\underline{\qquad} Z_1 \text{ X } Z_2 = r_1 \text{ X } r_2\ \Theta_1 + \Theta_2$

2. $\underline{\qquad} Z_1 \div Z_2 = r_1 \div r_2\ \Theta_1 - \Theta_2$

5.8 De Moivre's Theorem

$\underline{\qquad}$ Referring to section 5.7, $Z_1 = 10.44\ 73.3 = 3 + 10i\ (Z_1)^2 = (3 + 10i)^2 = (3 + 10i)(3 + 10i)$

$\underline{\qquad\qquad\qquad\qquad}(Z_1)^2 = (10.44\ 73.3\ )^2 = 10.44 \text{ X } 10.44\ 73.3 + 73.3 = 108.994\ 146.6 = (3 + 10i)^2$

$\underline{\qquad\qquad\qquad\qquad}(Z_1)^3 = (10.44\ 73.3\ )^3 = 10.44 \text{ X } 10.44 \text{ X} 10.44\ 73.3 + 73.3 + 73.3 = 108.994\ 146.6$

$\underline{\qquad\qquad\qquad}(Z_1)^4 = (10.44\ 73.3\ )^4 = 10.44 \text{ X } 10.44 \text{ X } 10.44 \text{ X } 10.44\ 73.3 + 73.3 + 73.3 + 73.3$

$\underline{\qquad} = 11879.605\ 293.2$

General Formula:

$\underline{\qquad}(Z_1)^n = (r_1\ \Theta_1\ )^n = (r_1\ )^n \Theta_1 \text{ X } n$

5.9 Complex Number Equations

Complex equations can be solved by equating the real components and the imaginary components of both sides of the equation.

Example 1 $x + 0.7yi = 2 + 4i$

$x = 2 \ 0.7y = 4 \ y = 4 \div 0.7 = 5.7143$

Example 2 $2x + 0.5y + ix + iy = 5 + 6i$

$2x + 0.5y + i(x + y) = 5 + 6i$

$2x + 0.5y = 5$ (real component)

$x + y = 6$ (imaginary component) $x = 6 - y$

$2(6 - y) + 0.5y = 5 \ 12 - 2y + 0.5y = 5 \ 12 - 1.5y = 5 \ y = 4.667$

$X = 6 - 4.667 = 1.333$

Check: $2(1.333) + 0.5(4.667) + i(1.333 + 4.667) = 4.9995 + 6i$

Example 3 $(3 + i)^8 = \dfrac{7x + i2y}{i}$

Let $3 + i = z_1 \ r_1 = (3^2 + 1^2)^{0.5} = 3.1623 \ \Theta_1 = \tan^{-1}\dfrac{1}{3} \ \Theta_1 = 18.4332^0$

Applying Demoivre's Theorem $(3 + i)^8 = (3.1623 \text{cis } 18.4332)^8 = 3.1623^8 \text{cis}(18.4332 \ X8) = 10\ 000.5652\text{cis}147.4656$

$10\ 000.5652\text{cis}147.4656 = -8431.1635 + 5378.3627i$

$\dfrac{7x + i2y}{i} \ X \ \dfrac{i}{i} = \dfrac{7ix + (-2y)}{-1} = 2y \ -7ix$ LHS = RHS. $2y \ -7ix = -8431.1635 + 5378.3627i$

$-7ix = 5378.3627i \ x = -768.3375$

$2y = -8431.1635 \ y = -4215.5818$

Check: $10\ 000.5652\text{cis}147.4656 =$

$\dfrac{7(-768.3375) + i2(-4215.5818)}{i} \ X \ \dfrac{i}{i}$

5.10 Powers of i Some complex equations may be given in which the powers of i has to be simplified. The evaluation of powers of is based on the even and odd powers as illustrated below. $y = x + iy$, $i^2 = -1$ (real component) $i^{21} = i(i^2)^{10} = i(-1)^{10} = i$ (imaginary component) $i^{20} = (i^2)^{10} = 1$ (real component) $i^{43} = i(i^2)^{21} = -i$ (imaginary component) $i^{42} = (i^2)^{21} = 1$ (real component)

Exercise 5.1 1. Evaluate the complex number addition and subtraction (Refer to Fig 5.3): a) $z_1 + z_2 + z_5$ b) $z_1 + z_2 + z_5 + z_6$ c) $z_1 + z_2 - z_5 + z_6 - z_3$ 2. a) Convert the following complex numbers in Fig 4.3 to polar Form. i) z_1 ii) z_2 iii) z_3 iv) z_4 v) z_5 vi) z_6 b) Evaluate and express in rectangular form i) $z_1 \times z_2$ ii) $z_4 \times z_5$ iii) $z^4{}_4 \times z_5$ iv) $z^4{}_4 \div z_5$ v) $(z_4 \times z_5)^4 \div z_6$ 3. a) Draw the following complex numbers in an argand diagram and convert to polar form. i) $z_1 = 4i$ ii) $z_2 = 8 - i$ iii) $z_3 = 6 + 6i$ iv) $z_4 = -3 - 9i$ v) $z_5 = 4 - 12i$ vi) $z_6 = 15$ b) Based on your polar complex numbers in Q3a evaluate the following. Leave your answers in polar form i) $z_3{}^2 \times z_4{}^6$ ii) $z_3{}^6 \div z_4{}^2$ iii) $z_6 \times z_2{}^3 \div z_5$ 4.Solve for x and y given the following complex equations. a) $8 - 15y = 3 + 20xi$ b) $10x + 4 - 12i + 18yi = 13 + 12i$ c) $x + 7 + i + 9yi = 17 + 13i$ d) $\dfrac{x+i8}{10+yi} = 20$

e) $(8 - 16i)^3 = x - iy$ f) $\dfrac{5y+10}{15+5i} = \dfrac{x+10yi}{11i+3}$

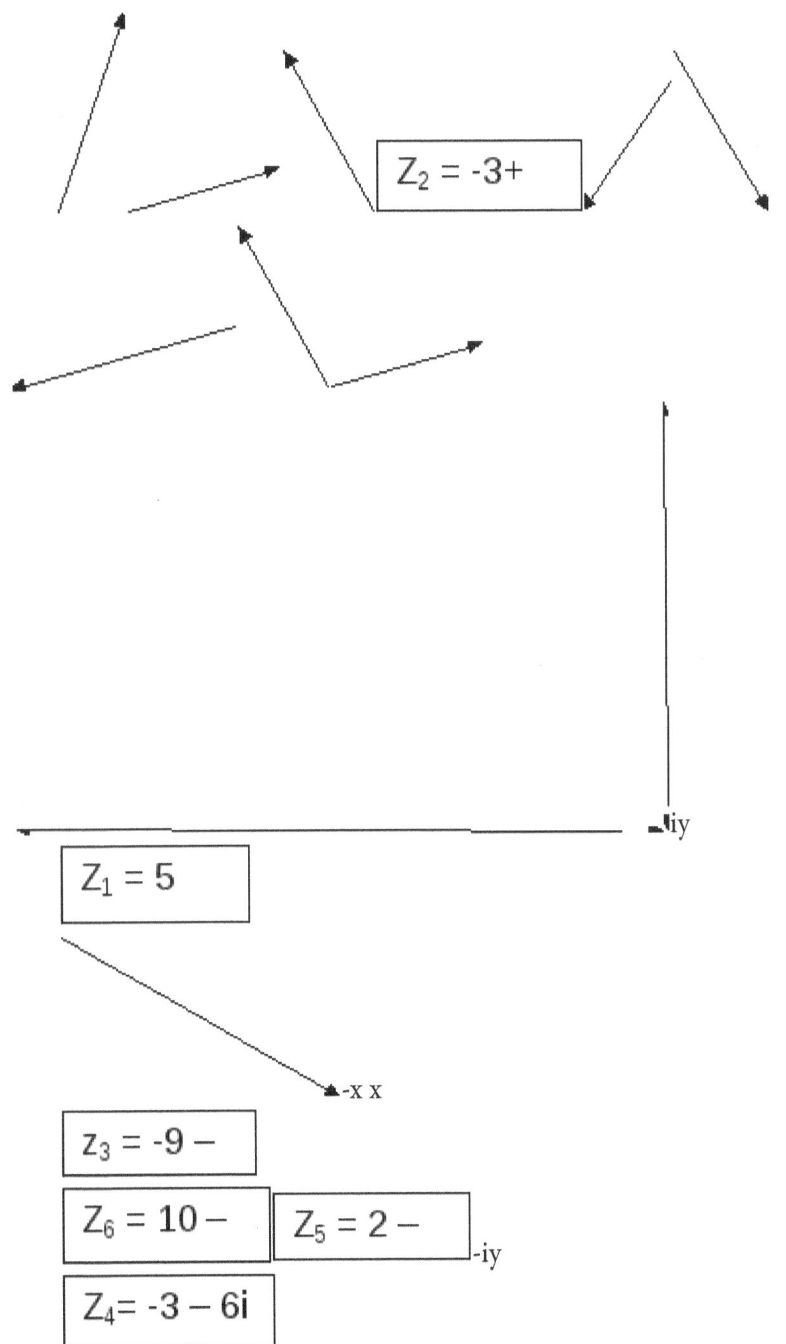

$Z_2 = -3+$
$Z_1 = 5$
+iy
-x x
$z_3 = -9 -$
$Z_6 = 10 -$
$Z_5 = 2 -$
-iy
$Z_4 = -3 - 6i$

Fig. 5.3 Argand Diagram

5.11 Application of Complex Numbers Complex Number calculations are applied in Engineering and Applied Sciences as follows.

Example 1 Calculate the resultant of the forces $z_1 = (5 + 2i)$ N and $z_2 = (-3 + 7i)$ N. What is the magnitude and direction of the resultant of these two forces? The resultant of the two forces, z_1 and z_2, is the addition of the vectors: $z_1 + z_2 = (5 + 2i) + (-3 + 7i) = 2 + 9i$ modulus, $r = (2^2 + 9^2)^{0.5} = 9.22$ Newtons argument, $\Theta = \tan^{-1} (9/2) = 77.47^0$ Resultant, $F = 9.22\ 77.47^0$N

Example 2. Calculate the resultant of the following forces acting at a point. a) A force of 350N at 30^0 b) A Force of 490N at 150^0 c) A force of 620N at 260^0 d) A force of 200N at 300^0

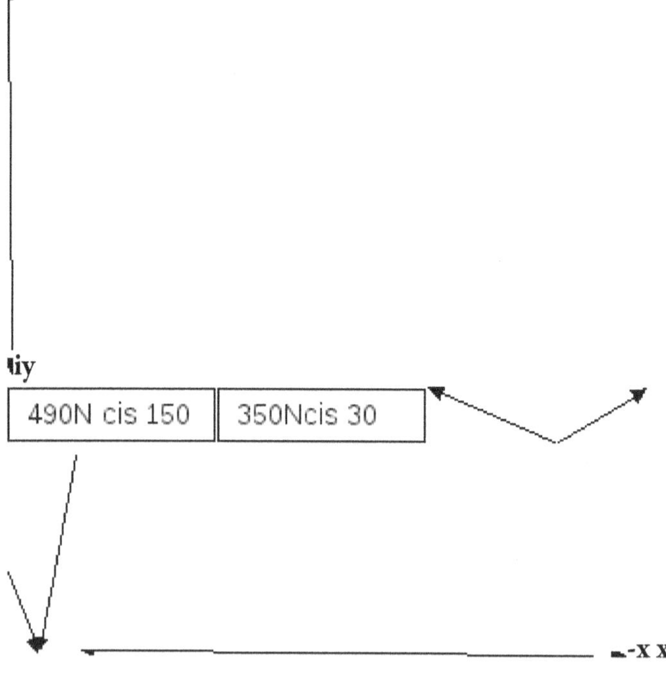

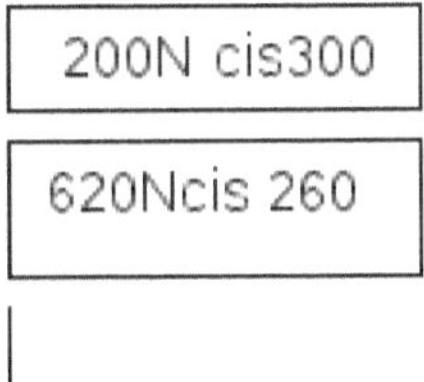

Fig 5.4 Calculating the resultant of a system of forces.

_______|_______ ____|___|___||____Resultant of forces: F = 350N 30 + 490N 150 + 620N 260 + 200N 300 F = 303.11 + 175i - 424.35 + 245i -107.66 -610.58i +100 -173.21i = -128.9 -363.79i

_____|F = 385.95 N at a positive angle of 250.49^0 = 385.95 250.49^0NExample 3 Fig 4.5 shows an RLC circuit diagram. By applying the given formulas, calculate the: a) the total impedance. b) The total current flowing through the circuit.

Total Impedance $Z_T = [Z_2Z_3 \div (Z_2 + Z_3)] + Z_1$ Total Current: $I_T = V_T/Z_T$

a. $Z_T = \dfrac{(5+9i)(6-12i)}{5+9i+6-12i} + (0-12i)$

$$\frac{10.296 \quad 60.95 \text{ X } 13.42 \quad 360\text{-}63.43}{11\text{-}3i} \quad \Bigg|_{-12i}$$

$$\frac{138.14 \quad 357.52}{11.4 \quad 360\text{-}15.26} - 12i = 12.12 \quad 12.78 - 12i$$

$=11.82 - 9.32i = 15.05$ 321.74 Ω

a. _______ ____ $I_T = \dfrac{200 \quad 0}{15.05 \quad 321.74} = 13.29$

-321.74 A

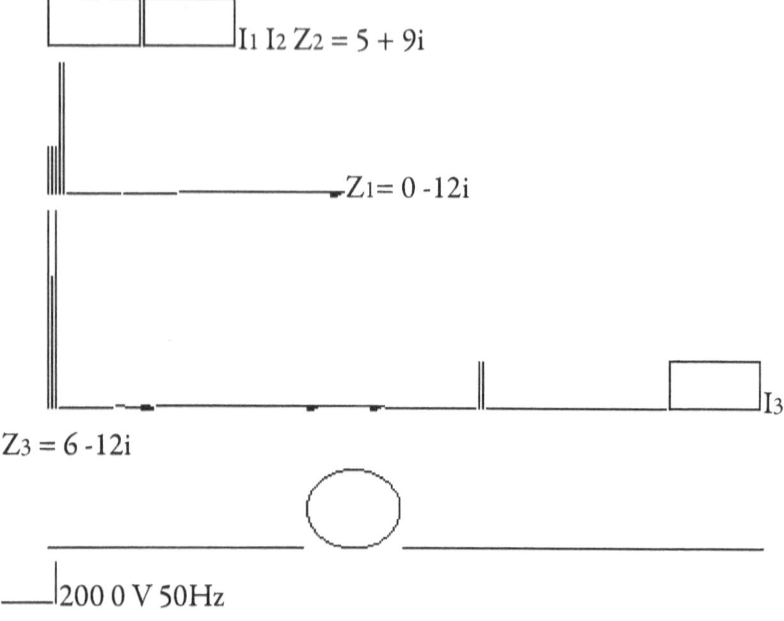

Fig 5.5 Calculation of Impedance and Total Current of RLC Circuit.

Example 4 Trapezium WXYZ is as shown in the argand diagram below. Calculate W, X, Y, and Z in rectangular and polar complex numbers. $X0 = 6$ units, $WZ = 4$ units, $Y = 6 + 3i$, $OZ = 4.025$ units and $OY = OP$

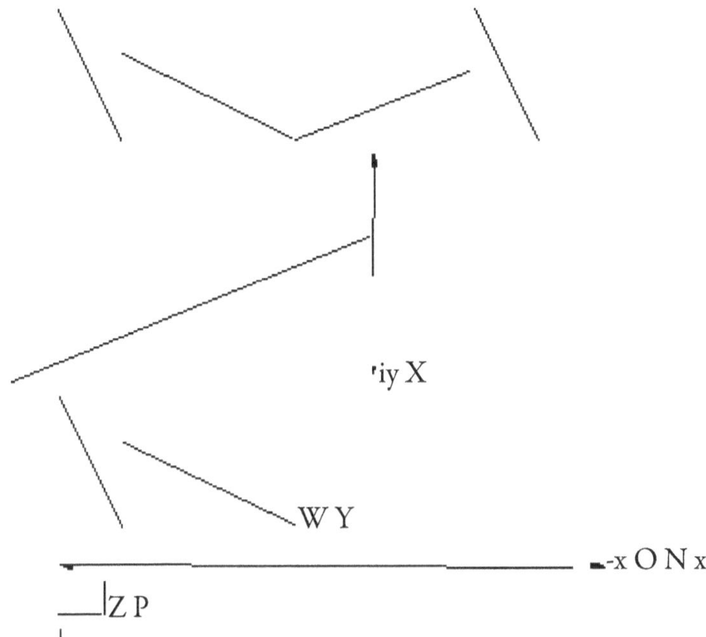

Fig 5.6 Argand Diagram of Trapezium WXYZ

Y = 6 + 3i = 6.71 26.57^0 X= 0 + 6i = 6 90^{0}Considering Parallelogram WOYX and Δ OYX: OY + YX = OX YX = OX – OY = (0 + 6i) – (6 + 3i) = -6 + 3i

W= YX -6 + 3i = 6.71 153.43 OZ = 4.025 = modulus = r Z = 4.025(cos (180 + 26.57) + isin (180 + 26.57) = 4.025 206.57 = -3.59 - i1.8

Exercise 5.2 1. Calculate the resultant of the system of forces acting on a point by firstly sketching the argand diagram of the system of forces: a) A force of 650N at 45^0 b) A Force of 190N at 190^{0}c) A force of 420N at 265^{0}d) A force of 350N at 290^0 2. Considering the circuit diagram in Example 3 above and applying the given formulas , calculate: a) the total impedance b) The total current flowing through the circuit given that Z_1= 0 -16i, Z_2 = 6 + 10i , Z_3 = 3 -18i and Voltage = 200 V at 50Hz. Total Impedance Z_T = [$Z_2Z_3 \div (Z_2 + Z_3)$]

+ Z_1 ; Total Current = I_1 = V_T/Z_T 3. A rectangle MNOP has the corner M lying on the point 0+0i. Given that NO = 4.5 units, MN = 1.6 units and the line PM is 135^0 with respect to the positive x axes. Draw and label MNOP in an argand diagram. Calculate the polar and rectangular complex numbers of the corners of the rectangle. 4. A rectangle WXYZ has the corner W lying on the point 2+2i. Given that WX = 4.5 units, WZ = 1.6 units and the line WX is 45^0 with respect to the positive x axes. Draw and label WXYZ in an argand diagram. Calculate the polar and rectangular complex numbers of the corners of the rectangle. 5. A plane flies in the north eastern direction at a velocity of 175 km/h. A wind of velocity 95 km/h and eastern direction affects the flight of the plane. Draw and label the resultant of the plane in an argand diagram. Calculate the magnitude and direction of the resultant of the plane.

Chapter 6 Differentiation

6.1 Introduction Differentiation is defined as the rate of small change in y with respect to small change in x. Differentiation may be defined from the trigonometric perspectives.

The slope of a straight line is m and m= $\tan \theta$ where $\tan \theta = \dfrac{\text{vertical distance}}{\text{horizontal distance}}$ .The slope of any given two pairs of points on a straight is a constant are both equal to m.

The slope of a tangent passing through any two given points on a curve is not a constant. Differentiation is usually applied in calculating the slope of a tangent passing through a point on a curve as shown in Example 2. Mathematically, $\dfrac{dy}{dx} = \dfrac{\text{small change in y}}{\text{small change in x}} = \dfrac{\Delta y}{\Delta x}$

6.2 Standard Formula for Differentiation. For any given expression $y = ax^n$, the differential of ax^n is: $\dfrac{dy}{dx} = nax^{n-1}$, n = power to which x is raised a= coefficient of x.

Example 1 Calculate the differential of $6x^2 + 17x + 5$. The derivative of the above quadratic equation is calculated by applying the standard formula to each term. The differential of $6x^2 + 17x + 5$ is the sum of the derivative of each term. $\dfrac{dy}{dx} = 12x + 17$

6.3 Differentiation: increasing and decreasing functions The graphs and tables, Fig 6.2a, Fig 6.2b, Table 6.2a and Table 6.2b, respectively will be our reference for illustrating increasing and decreasing functions. Given that, Resistance, R, is a constant and applying the formula for Voltage, V = IR, the values of table 6.1 have been completed as shown. Similarly, Torque, T = FR where F = Force and R= Radius.

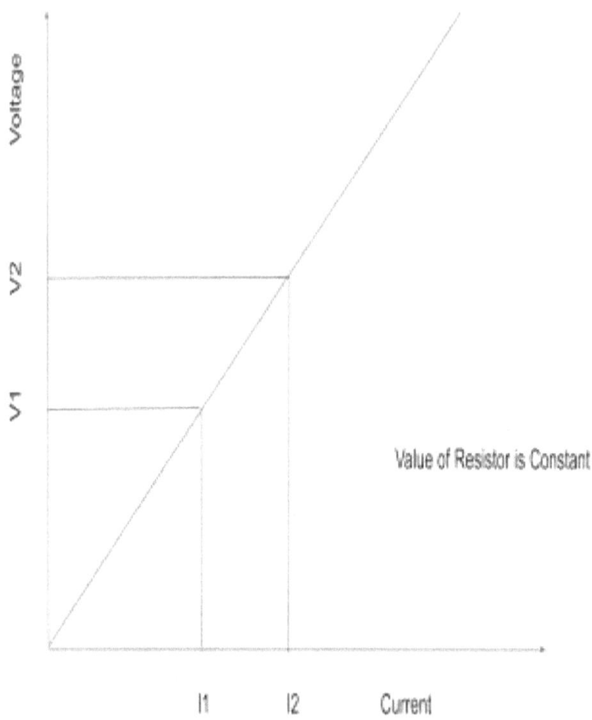

Fig 6.1 Increasing Function

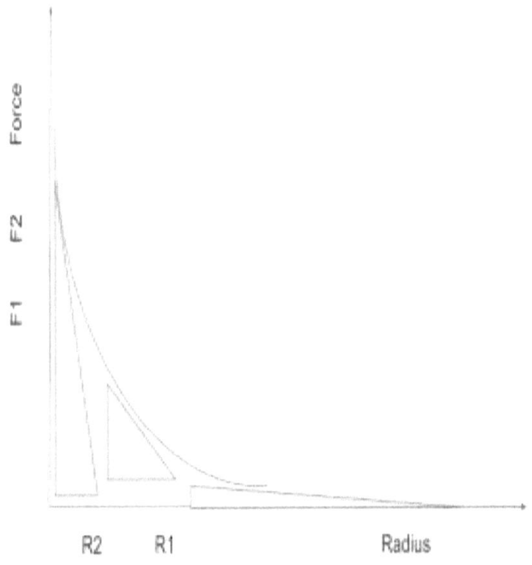

Fig. 6.2 Decreasing Function. The Hyperbolic Function T=FR.

An increasing function is a function in which an increase of voltage from V_1 to V_2 corresponds to an increase in current from I_1 to I_2. This function is also known as direct proportionality. The differential of an increasing function is a positive value. A decreasing function is a function whereby an increase in force from F_1 to F_2 corresponds to a decrease in radius from R_1 to R_2 respectively. This function is referred to as indirect proportionality. The differential of a decreasing function is a negative value. The derivative of the curve at any of the points A or B is negative. This means that the slope of the tangents passing through points A and B are negative. The derivatives of increasing and decreasing functions may also be described from the trigonometric perspective thus. The plotted line in the graph of table 6.1 is an acute angle and the tan of an acute angle is positive. On the other hand, the tangents passing through the points A and B as shown in the graph (Fig 6.2) makes an angle in the second quadrant. The tan of an angle in the second quadrant is negative. Generally, it is very important to understand the relationship between letters and symbols in engineering calculations. This is relevant as we are able confirm or verify our calculations based on this relationship. In conclusion, the table below summarizes increasing and decreasing functions.

Table 6.1 Increasing function

Constant Resistance (Ohms)	Volts (V)	Current A (Amperes)
2	24	12
2	12	6
2	8	4
2	4	2
2	2	1
2	1	0.5

Table 6.2 Decreasing function

Constant

Torque (Nm)	Force F(Newton)	Radius R (meters
25	25	1
25	20	1.25
25	12	2.083
25	10	2.5
25	5	5
25	2	12.5

Table 6.3

	Increasing Function	Decreasing Function	Derivative
$y = mx + c$	$y = 4x + 2$		4
$y = -mx + c$ where $k = m$ $m = y/x$		$y = -2x + 5$	-2
$K = xy$		$2 = xy$	$-2x^{-2}$
$y = x^2$	$y = x^2$ $0 \leq x \leq +\infty$	$y = x^2$ $-\infty \leq x \leq 0$	$2x$

6.4 Differentiation: Rates of Change See Section 6.10

6.5 Differentiation: From First Principles

Differentiation from first principles is based on the formula as stated below in Equation 1.

$$\frac{dy}{dx} = \lim_{h \to 0} \frac{f(h+x)-f(x)}{h} \quad (1)$$

$$\frac{dy}{dx} = \frac{\Delta y}{\Delta x} \quad (2)$$

Since both Equ. 1 and 2 are equations of differentiation, it would be logical to compare the right hand side of each equation. Comparing,

we have: $\dfrac{f(h+x)-f(x)}{h} = \dfrac{\Delta y}{\Delta x}$

Therefore, the term, $f(h+x) - f(x)$, is equal to the vertical component while the term h is equivalent to the horizontal component. In addition to the above, h tends to 0 refers to x value of the turning point or the line of symmetry of the curve. This line makes an angle of 90 with the horizontal and it does not have a horizontal component.

Example 2 Find $\dfrac{dy}{dx}$ of $y = 3x - 6$. Applying the standard form for differentiation to $3x^1$ we have $(1)3x^{1-1} = 3x^0 = 3$. Compare the straight line equation with the standard form $y = mx + c$.

The differential of a straight line = tan Θ = slope slope = gradient = m. $\Theta = \tan^{-1} 3 = 71.5650$ [$\dfrac{d(6)}{dx} = 0$ since 6 does not have an x term]

Example 3 Find the gradient of the quadratic curve $y = 6x^2 + 11x + 3$ at the points a) x, y = (0, 3) b) x, y = (-0.917, -2.042) c) x, y = (-1.5, 0)

Solution:

a. $\dfrac{dy}{dx} = 12x + 11$ Substitute values of $x = 0$ into $12x + 11$

$12(0) + 11 = 11$ (positive slope)

Tangent passing through point (0,3) on curve makes an angle of $\Theta = \tan^{-1} 11 = 84.81^0$ with positive x axis. b) x = -0.917 $\dfrac{dy}{dx} = 12(-0.917) + 11 = 0$

Tangent passing through point (-0.917, -2.042) on curve makes an angle of

$\Theta = \tan^{-1}(0) = 0$ with positive x axis. c) $X = -1.5$ $\frac{dy}{dx} =$ $12(-1.5) + 11 = -7$ (negative slope) Tangent passing through point (-1.5, 0) on curve makes an angle of $\Theta = \tan^{-1}-7 = 98.13^0$ with positive x axis.

The above examples are applications of differentiation on straight lines and quadratic equations respectively.

6.6 Methods of Differentiation 6.6.1 Product Rule The

product rule states that given an equation in the standard form

a. $y = uv$, $\frac{dy}{dx} = u\frac{dv}{dx} + v\frac{du}{dx}$

b. $y = uvw$, $\frac{dy}{dx} = vw\frac{du}{dx} + uw\frac{dv}{dx} + uv\frac{dw}{dx}$

The two formulas above can be explained thus: a) Given a product of two functions, uv, the differential is the sum of: i) the product of u and the derivative of v ii) the product of v and the differential of u b) Given a product of three functions, uvw, the derivative is the sum of: i) the product of v, w and the derivative of u ii) the product of u, w and the differential of v iii) the product of u, v and the derivative of w

Example 1 Find $\frac{dy}{dx}$, given that $y = (x^2 + 8x + 7).\sin x$ Let $u = x^2 + 8x + 7$ and $v = \sin x$ $\frac{dy}{dx} = (x^2 + 8x + 7).\frac{d(\sin x)}{dx} + (\sin x)\frac{d(x^2+8x+7))}{dx} =$ $(x^2 + 8x + 7).\cos x + (2x + 8)\sin x$

Example 2 Find the differential of $y = x^3 (\sin x) (2x - 10)$.

Let $u = x^3$, $v = \sin x$ and $w = 2x - 10$

$$\frac{dy}{dx} = (\sin x)(2x\text{-}10)\frac{d(x^3)}{dx} + x^3.(2x\text{-}10)\frac{d(\sin x)}{dx} + x^3.(\sin x)\frac{d(2x\text{-}10)}{dx}$$

$$= (\sin x)(2x\text{ -}10)3x^2 + x^3.(2x\text{-}10)\cos x + 2x^3.(\sin x) = x^2[3(2x\text{-}10)\sin x + x(2x - 10)\cos x + 2x\sin x]$$

Example 3 Given that $y = x^3 e^x(\cos x)$, calculate $\frac{dy}{dx}$.

Let $u = x^3$, $v = e^x$ and $w = \cos x$;

$$\frac{dy}{dx} = e^x.\cos x\frac{d(x^3)}{dx} + x^3.\cos x\frac{d(e^x)}{dx} + x^3.e^x\frac{d(\cos x)}{dx}$$

$$= 3x^2.e^x.\cos x + x^3.e^x.\cos x + x^3.e^x.(\text{-}\sin x) = e^x[(3x^2 + x^3)\cos x - x^3\sin x]$$

6.6.2 Quotient Rule The Quotient rule states that given an equation of the form:

$$y = \frac{u}{v}, \quad \frac{dy}{dx} = \frac{v\frac{du}{dx} - u\frac{dv}{dx}}{v^2}$$

In this case, a problem is expressed in the form of a fraction where there is a numerator and a denominator. As stated above, the numerator is u and the denominator is v. The differential, $\frac{dy}{dx}$, is explained as follows. a) The numerator is the difference of : i) the product of v and the derivative of u ii) the product of u and the differential of v b) the denominator is the square of the v. **Example 1** Evaluate $\frac{dy}{dx}$ of $y = \frac{x}{3\text{-}x^3}$

$$U = x \text{ and } v = 3 - x^3$$

$$\frac{dy}{dx} = \frac{(3 - x^3)\frac{d(x)}{dx} - (x)\frac{d(3 - x^3)}{dx}}{(3 - x^3)^2}$$

$$= \frac{(3\text{-}x^3)(1) - (x)(\text{-}3x^2)}{(3\text{-}x^3)^2} \quad = \frac{3 + 2x^3}{(3\text{-}x^3)^2}$$

Example 2 Evaluate $\dfrac{dy}{dx}$ of $y = \tan x$

$$y = \frac{\sin x}{\cos x} = \frac{u}{v}$$

$$\frac{dy}{dx} = \frac{(\cos x)\dfrac{d(\sin x)}{dx} - (\sin x)\dfrac{d(\cos x)}{dx}}{(\cos x)^2}$$

$$= \frac{(\cos x)(\cos x) - (\sin x)(-\sin x)}{(\cos x)^2} = \frac{(\cos^2 x) + (\sin^2 x)}{(\cos x)^2}$$

$$= \frac{1}{(\cos x)^2} = \sec^2 x \ [\cos^2 x + \sin^2 x = 1]$$

Example 3 Find $\dfrac{dy}{dx}$ of $y = \dfrac{e^x}{\sin x}$

$$= \frac{(\sin x)\dfrac{d(e^x)}{dx} - (e^x)\dfrac{d(\sin x)}{dx}}{(\sin x)^2} = \frac{(e^x)(\sin x - \cos\cos x)}{(\sin x)^2}$$

6.6.3 Function of a Function (Chain Rule)

Consider the following equations that have two functions:

a. $y = (x^3 + 10x - 15)^9$: cubic function raised to the power 9 (exponential function).

b. $y = \sin(2x^2)$: the sine of quadratic function.

c. $y = e^{4x}$: e raised to the power 4x as against to the power x (e^x). Exponential function raised to a linear function.

Compare the above equations with the previous examples for product and quotient rule. The evaluation of these differentiation problems are based on a substitution that reduces the problem into a standard or basic derivative as stated in tables.

The above equations can easily be solved by applying the function of a function formula or chain rule. For $y = (x^3 + 10x - 15)^9$ $y = u^9$

where $u = x^3 + 10x - 15$. The general formula for chain rule is $\frac{dy}{dx} = \frac{dy}{du} \times \frac{du}{dx}$.

Example 1 Find $\frac{dy}{dx}$ of $y = (x^3 + 10x - 15)^9$. Substituting $u = x^3 + 10x -15$ $y = u^9$ $\frac{dy}{dx} = \frac{dy}{du} \times \frac{du}{dx}$

$\frac{dy}{dx} = \frac{d}{du}(u^9) \times \frac{d}{dx}(x^3 + 10x - 15) = 9u^8 \times (3x^2 + 10)$ Substituting $u = x^3 + 10x -15 = 9(x^3 + 10x - 15)^8(3x^2 + 10)$.

Careful consideration should be given to the fact that function of a function is the product of two differentials. The substitution $u = x^3 + 10x -15$ shows the relationship between y and x. In addition to this, it simplifies the differentiation hence the name function of a function.

Example 2 Find $\frac{dy}{dx}$ of $y = \sin(2x^2)$. Let $u = 2x^2$ $y = \sin u$ $\frac{dy}{dx} = \frac{dy}{du} \times \frac{du}{dx}$ $\frac{dy}{dx} = \frac{d}{du}(\sin u) \times \frac{d}{dx}(2x^2) = 4x\cos u = 4x\cos(2x^2)$

Example 3 Given that $y = e^{4x}$, find $\frac{dy}{dx}$.

Let $u = 4x$ and $y = e^u$ $\frac{dy}{dx} = \frac{dy}{du} \times \frac{du}{dx}$ $\frac{dy}{dx} = \frac{d}{du}(e^u) \times \frac{d}{dx}(4x) = 4e^{4x}$

The concept of function of function may also be understood by comparing with the most basic corresponding functions. Most chain rule problems are derived by replacing the x which is a linear function with another standard function such as cos x, 2x, $3x^2 + 2x +10$ and x^3 etc. This is true for the rows 1, 2, 3 and 9 for the table 6.4 below.

Furthermore, the function of a function problems in row 4, 5 and 12, were derived by substituting the power of 1 in the second column with 10, In x and sinx respectively. Considering the functions $y = \text{cosec } x$, $y = (\text{cosec } x)^2$ and $y = \text{cosec}^2 3x^3$, the first function is the basic function. For the second function, it is clear that $y = \text{cosec } x$ has been raised to the power 2. Finally, x has been replaced by $3x^3$ in the second function to obtain the function of a function problem.

Table 6.4 Function of a function

Standard Functions	Standard Functions in terms of $f(x)$ or raised to power 1 $f(x) = x$	Chain Rule Problems
1 $y = e^x$	$y = e^{f(x)}$	$y = e^{\cos x}$
2 $y = a^x$	$y = a^{f(x)}$	$y = a^{2x}$
3 $y = \ln x$	$y = \ln f(x)$	$y = \ln (3x^2 + 2x + 10)$
4 $y = x^2 + 2x + \dfrac{4}{4}$	$y = (x^2 + 2x + 4)^1$	$y = (x^2 + 2x + 4)^{10}$
5 $y = 2x + 9$	$y = (2x + 9)^1$	$y = (2x + 9)^{\ln x}$
6 $y = \sin x$	$y = \sin f(x)$	$y = \sin 3x$
7 $y = \cos x$	$y = \cos f(x)$	$y = \cos \ln x$
8 $y = \tan x$	$y = \tan f(x)$	$y = \tan^2 x$
9 $y = \text{cosec } x$	$y = \text{cosec } f(x)$	$y = \text{cosec } x^3$
10 $y = \sec x$	$y = \sec f(x)$	$y = \sec (x^3 + 2x^2 + 6x)$
11 $y = \cot x$	$y = \cot f(x)$	$y = \cot x^3$
12 $y = x^3$	$Y = (x^3)^1$	$y = (x^3)^{\sin x}$

These problems may also be solved by comparing with the corresponding standard differentials formula from differentiation tables. (Table 6.5)

Table 6.5 Table of Standard Differentials

y	$\dfrac{dy}{dx}$
1. ax^n	nax^{n-1}
2. e^x	e^x
3. $Inx = \log_e x$	$\dfrac{1}{x}$
4. a^x	$a^x.\log_e a$
5. $sinx$	$cosx$
6. $cosx$	$-sinx$
7. $tanx$	$sec^2 x$
8. $cosecx$	$-cosec\ x.tanx$
9. $secx$	$sec\ x.tanx$
10. $cot\ x$	$-cosec^2 x$

Exercise 6.1

1. Differentiate the following functions by applying the standard formula for differentiation $\dfrac{dy}{dx} = nax^{n-1}$.

    a. $y = x^3 + 3x^2 - 16x - 48$

    b. $y = 3x^3 + 7x^2 + 8$

    c. $y = x^3 + 5x^2 - x - 5$

1. Calculate $\dfrac{dy}{dx}$ of the following:

a. $y = \tan x$

b. $y = \cot x$

c. $y = \csc x$

d. $y = \sec x$

e. $y = \dfrac{1}{2x^3 - 3x}$

f. $y = (3x + 10)^3$

1. Find $\dfrac{dy}{dx}$ by applying an appropriate differentiation method:

a. $y = 6\sin 7x$

b. $y = 10x^4 . e^x$

c. $y = \dfrac{2x^3}{\cos x}$

d. $y = (\cos x)^6$

e. $y = (x^3 + 3x^2 - 16x - 48)^{21}$

f. $y = 10e^{8x}$

6.7 Differentiation of Trigonometric Functions

Trigonometric equations have to be simplified in order to be differentiated. The following examples illustrate this.

Example 1 Find $\dfrac{dy}{dx}$, given that $y = \dfrac{\cos 2x}{\cos^2 x}$.

$$y = \dfrac{\cos^2 x - \sin^2 x}{\cos^2 x} = 1 - \tan^2 x$$

$$\frac{dy}{dx} \text{ of } 1 = 0$$

$\frac{d}{dx}(-\tan^2 x)$ applying function of a function

Let $u = \tan x$ and $y = 1 - \tan^2 x$ $y = 1 - u^2$ $\frac{dy}{dx} = \frac{dy}{du} \times \frac{du}{dx}$

$\frac{dy}{dx} = \frac{d}{du}(1 - u^2) \times \frac{d}{dx}(\tan x) = -2u \times \sec^2 x = -2\tan x \sec^2 x.$

This problem may also be evaluated by applying the quotient rule or product rule.

Example 2. Evaluate the Differential of $y = \frac{\cos^2 x + \sin^2 x}{\cosec^2 x}$

$y = \frac{1}{\cosec^2 x} = \sin^2 x$

$y = u^2$ Let $u = \sin x$ $\frac{dy}{dx} = \frac{dy}{du} \times \frac{du}{dx}$ $\frac{dy}{dx} = \frac{d}{du}(u^2) \times \frac{d}{dx}(\sin x) = 2u \times \cos x = 2\sin x \cos x$

Example 3 Given that $y = \frac{10}{\cot\theta\sec\theta}$, calculate $\frac{dy}{d\theta}$.

$y = 10\tan\theta\cos\theta$

Let $u = \tan\theta$ and $v = \cos\theta$

$$\frac{dy}{d\theta} = 10\frac{d}{d\theta}[\tan^\theta \cos\theta]$$

$$= \tan^\theta \frac{d}{d\theta}[\cos^\theta] + \cos^\theta \frac{d}{d\theta}[\tan\theta]$$

$$= 10[\tan^\theta[-\sin^\theta] + \cos\theta\sec 2\theta] = 10[\cos\theta\sec 2\theta - \tan^\theta\sin^\theta]$$

Example 4 Evaluate $\frac{dy}{dA}$, if $y = \dfrac{2\cos 2A - 1 - \cos 2A - 1}{\cos A}$.

$$y = \frac{2\cos^2 A - 1 - 2}{\cos A} = \frac{2\cos^2 A - 3}{\cos A} = 2\cos A - \frac{3}{\cos A}$$

applying quotient rule to differentiate $\dfrac{3}{\cos A}$, we have

$$\frac{d}{dA}\left[\frac{3}{\cos A}\right] = \frac{(\cos A)\frac{d(3)}{dA} - (2)\frac{d(\cos A)}{dA}}{(\cos A)^2} = \frac{(\cos A)(0) - (3)(-\sin A)}{(\cos A)^2}$$

$$= \frac{3\sin A}{\cos^2 A}\ \frac{d}{dA}\left[2\cos A - \frac{2}{\cos A}\right] = -2\sin A + \frac{3\sin A}{\cos^2 A} =$$

$$-2\sin A + \frac{3\tan A}{\cos A}.$$

Applying function of a function on $y = 2\cos A - 3(\cos A)^{-1}$ will give the same answer.

Exercise 6.2

Differentiate the following equations.

1. $y = \sin^2 x + \sin^4 x + \cos^2 x$

2. $y = \dfrac{\cot^3 A}{1 - \csc^2 A}$

3. $y = \dfrac{\tan^5 A}{1 - \sec^2 A}$

4. $y = \sin^2 x - 6\sin x + 9$

5. $y = 9 - \cot^2 x$

6. $y = \dfrac{10\cos 2\theta}{\sin 2\theta} \cdot (\cot^2 \theta + 1)$

7. $y = (\cot^2 A + \csc^2 A)\sin^2 A$

8. $y = (\tan^2 A + \sec^2 A)\cos^2 A$

9. $y = \dfrac{15\cos 3\theta}{\sin 3\theta} \cdot (\tan^2 \theta + 1)\cos^2 \theta$

10. $y = [1 + \cot^2 A]^4$

11. $y = [\csc^2 x - 1]^{0.5}$

12. $y = [\csc^2 x - 1]^2$

6.8 Differentiation of Logarithmic and Exponential Functions

Logarithmic and Exponential functions can be differentiated by applying the formula as stated in table 6.5.

a. Given that $y = \ln x$ $dy/dx = 1/x$.

b. $y = e^x$ $dy/dx = e^x$

c. $y = 10^x$ $dy/dx = 10^x \ln 10$

However, we may apply function of a function based on the given problem.

Example 1 Find the differential of $y = 4e^{4x}$.

Let $u = 4x$, $y = 4e^u$

$\dfrac{dy}{dx} = \dfrac{dy}{du} \times \dfrac{du}{dx} = 4e^u \cdot 4 = 16e^{4x}$

Example 2 Evaluate $\dfrac{dy}{dx}$ of $y = \ln 4x^2$

Let $u = 4x^2$ and $y = \ln u$

$$\dfrac{dy}{dx} = \dfrac{dy}{du} \times \dfrac{du}{dx} = \dfrac{1}{u} \times 4(2x) = \dfrac{8x}{4x^2} = \dfrac{2}{x}$$

Example 3 Find the derivative of $y = 3e^{-3x} - 18^x + 200\ln x$

$$\dfrac{dy}{dx} = -3(3e^{-3x}) - 18^x \ln 18 + \dfrac{200}{x} = -9e^{-3x} - 18^x \ln 18 + \dfrac{200}{x}$$

Exercise 6.3 Calculate the derivative of: 1. $y = \log 10x^3$ 2. $y = 60e^{2x}$ 3. $y = 5\ln 6x^4$ 4. $y = 25e^{3x} + 5\ln 6x^5$ 5. $y = 10\ln \sqrt{8x^2}$ 6. $y = 2e^{4x} - 8\log 7x^{10}$

6.9 Binomial Theorem

Consider the multiplication of the following:

   a. $(x + 3)(x + 3) = x^2 + 6x + 9$

   b. $(x + 3)^{1/2}$

The expansion of $(x+3)^{1/2}$ can be carried out by applying the Binomial Theorem which states that

$$(x + y)^n = x^n + \dfrac{n\, x^{n-1}y^1}{1\times1} + \dfrac{n(n-1)\, x^{n-2}y^2}{1\times2} + \dfrac{n(n-1)(n-2)\, x^{n-3}y^3}{1\times2\times3} + \dfrac{n(n-1)(n-2)(n-3)\, x^{n-4}y^4}{1\times2\times3\times4} \quad \ldots\ldots$$

Example 1 Evaluate $(x + 4)^4$

Comparing $(x + 4)^4$ and $(x + y)^n$: $y = 4$ and $n = 4$

Substituting into the Binomial theorem formula:

$$x^4 + \dfrac{4\, x^{4-1}4^1}{1\times1} + \dfrac{4(4-1)\, x^{4-2}4^2}{1\times2} + \dfrac{4(4-1)(4-2)x^{4-3}4^3}{1\times2\times3} + \dfrac{4(4-1)(4-2)(4-3)x^{4-4}4^4}{1\times2\times3\times4}$$

$$= x^4 + 16x^3 + 96x^2 + 256x + 256$$

Confirm the above expression by expanding $(x+4)^2(x+4)^2$ There is a definite number of terms after expansion. In this case, 5 terms. This is not true for the expansion of $(x + y)^{1/n}$.

Example 2 Expand $(x+3)^{1/3}$

$y = 3$ and $n = 1/3$

$$x^{1/3} + \frac{\frac{1}{3}x^{(1/3)-1}3^1}{1\text{X}1} + \frac{\frac{1}{3}(\frac{1}{3}-1)\,x^{\frac{1}{3}-2}3^2}{1\text{X}2} + \frac{\frac{1}{3}(\frac{1}{3}-1)(\frac{1}{3}-2)\,x^{\frac{1}{3}-3}3^3}{1\text{X}2\text{X}3} +$$

$$\frac{\frac{1}{3}(\frac{1}{3}-1)(\frac{1}{3}-2)(\frac{1}{3}-3)\,x^{\frac{1}{3}-4}3^4}{1\text{X}2\text{X}3\text{X}4}$$

$$= x^{1/3} + \frac{x^{-2/3}3^1}{3} + \frac{\frac{-2}{9}x^{\frac{-5}{3}}3^2}{1\text{X}2} + \frac{\frac{-10}{27}x^{-\frac{8}{3}}3^3}{1\text{X}2\text{X}3} + \frac{\frac{1}{3}(\frac{-2}{3})(\frac{-5}{3})(\frac{-8}{3})\,x^{\frac{-11}{3}}3^4}{1\text{X}2\text{X}3\text{X}4} + \ldots$$

$$= x^{1/3} + x^{-2/3} - x^{-5/3} + \frac{5}{3}x^{-8/3} + \ldots$$

The expansion of $(x +a)^n$ where n is less than 1 is an infinite series. This is because the number of terms after expansion is not definite.

Compare the final solution of the expansion of $(x +3)^3$ and $(x+3)^{1/3}$ by expanding $(x+3)^{1/3}$ to 8th term. The expansion of $(x - y)^n$ should be written in the form $(x + (-y))^n$ before the formula is applied.

Exercise 6.4

Expand the following by applying Binomial Theorem. Expand to the fourth term when n<1. 1. $(2x + 1)^{1/4}$ 2. $(x - 3)^{1/3}$ 3. $(3x - 2)^4$ 4. $(x - 2)^{1/5}$ 5. $(3x + 1)^5$ 6. $(x - 4)^5$

1. **Applications of Differentiation 6.10.1 Maximum and Minimum Points.** The maximum and minimum of a curve is given by the formula: 1.$\frac{dy}{dx} = 0$. The slope of the tangent passing through this

point is equal to zero. $\text{Tan}^{-1} 0 = 0^0$. 2. The point of inflexion is evaluated by the formula $\frac{d^2y}{dx^2} = 0$ This is the differential of the derivative of the given function. 3. The maximum and minimum points are calculated by substituting the x value of the maximum or minimum point as follows: a) minimum point: $\frac{d^2y}{dx^2} =$ positive value b) maximum point : $\frac{d^2y}{dx^2} =$ negative value It is usually necessary to complete a table of values and draw the curve based on the calculated points as shown below.

Example 1. Sketch the cubic curve $y = x^3 - 9x^2 + 18x$ by calculating the following: a) the maximum and minimum point b) the point of inflexion c) the roots of the cubic graph d) the y intercept.

Solution: a)$\frac{dy}{dx} = 3x^2 -18x + 18 = 0$ $3(x^2 -6x + 6) = 0$

therefore the roots x = $\dfrac{-(-6) \pm \sqrt{(-6)^2 - 4(1)(6)}}{2(1)}$

$X_1 = 4.7321$ and $x_2 = 1.2679$ $f(x_1) = -10.3923$; $f(x_2) = 10.3923$

$\frac{d^2y}{dx^2} = 6x -18$ hence $f(x_1) = 10.3923$ (minimum) $f(x_2) = -10.3923$ (maximum)

a. $\frac{d^2y}{dx^2} = 6x -18 = 0$ $x = 3$ and $f(3) = 0$
b. By applying the remainder theorem, $x = 3, x = 6$ and $x = 0$

$f(0) = 0$ and y intercept $= 0$.

Table 6.6 $y = x^3 - 9x^2 + 18x$

X -2 0 1.2679 3 4.7321 6 7

Y -80 0 10.3923 0 -10.392 0 28

Example 2 Sketch and label the graph of $y = x^3 - 9x^2 + 26x - 24$ by evaluating the following: a) the maximum and minimum points b) the point of inflexion c) the roots d) the y intercept Solution: Refer to

Table 6.7 $\dfrac{dy}{dx} = 3x^2 - 18x + 26$

$= 0$ the roots : $x = \dfrac{-(-18) \pm \sqrt{(-18)^2 - 4(3)(26)}}{2(3)}$ $x_1 = 3.57$ $x_2 = 2.423$

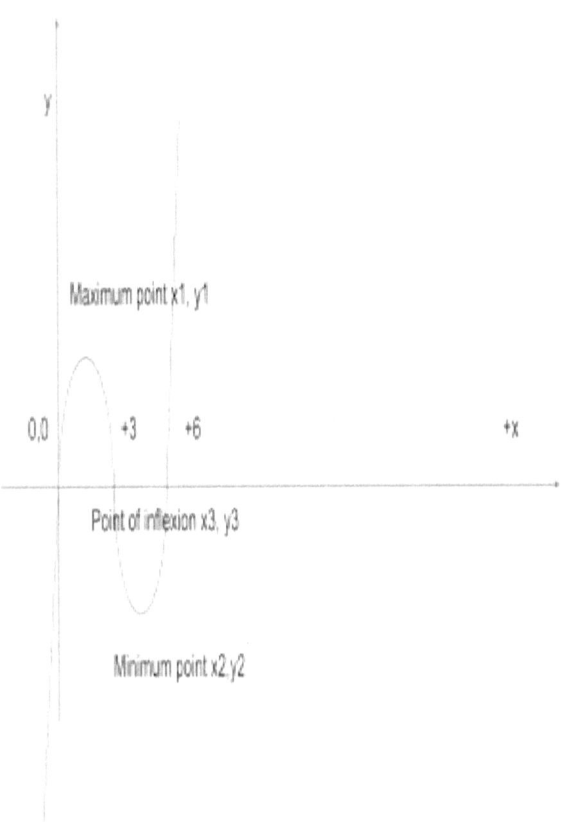

Fig 6.3 Cubic curve $y = x^3 - 9x^2 + 18x$

$$\frac{d^2y}{dx^2} = 6x - 18;$$ $f(x_1) = 3.462$ (minimum) and $f(x_2) = -3.462$ (maximum).

$X_1, y_1 = (3.577, -0.385)$ $x_2, y_2 = (2.423, 0.385)$

a. $\dfrac{d^2y}{dx^2} = 6x - 18 = 0$ $x = 3$ and $f(x) = 0$ therefore $x, y = (0,3)$

b. $y = x^3 - 9x^2 + 26x - 24 = (x-3)(x-4)(x-2)$ hence x=3, x= 4, x = 2 when y=0

$f(0) = -24$; x,y = (0, -24)

Table 6.7 $y = x^3 - 9x^2 + 26x - 24$

X0   22.4*   2.5   33.5 *   3.6   45

Y -240 0.385* 0.375 0 -0.375* -0.385 0 6

Example 3 Given the graph of y= 4cos4(x +30) evaluate the turning points and the point of inflexion of the trigonometric curve. Refer to Fig 1.20.

a. $\dfrac{dy}{dx} = \dfrac{dy}{du} \times \dfrac{du}{dx} = 4\left[\dfrac{d\,(\cos u)}{du} \times \dfrac{d(4(x+30)}{dx}\right] = 4[-\sin u \times 4] = -16\sin 4(x+30)$

Where u = 4(x + 30) and y = 4cos u

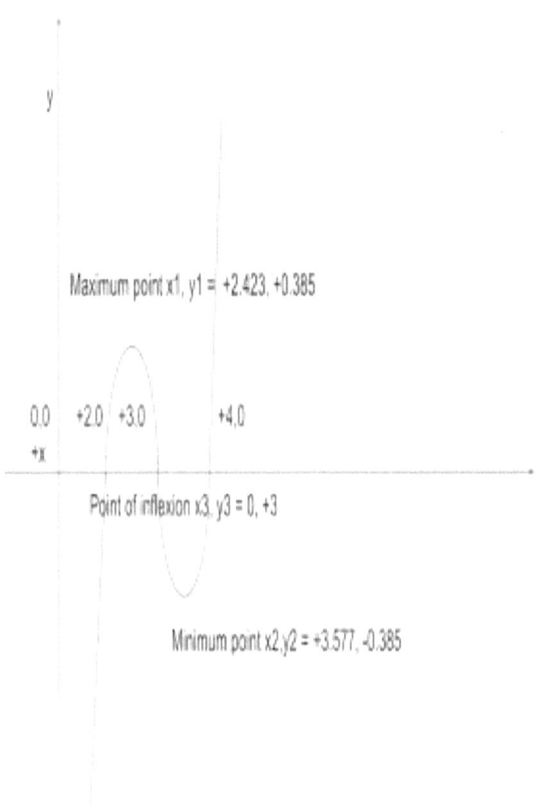

Fig 6.4 Graph of $y = x^3 - 9x^2 + 26x - 24$

$-16\sin 4(x+30) = 0$ $4(x+30) = \sin^{-1}(0)$ $0 = 4(x+30)$ $x = -30^0$ or 330^0 and $y = 4$

Remembering that 1 cycle $= 90^0$ and reading off each preceding maximum point from the graph, we have $(x, y) = $ 330, 4; 150, 4; 240; 4, 60, 4

Minimum Points: $x,y = 285, -4$; $195, -4$; $105, -4$; $15, -4$.

a. Point of inflexion $\dfrac{d^2y}{dx^2} = -16\left[\dfrac{d(\sin u)}{du} \times \dfrac{d(4(x+30)}{dx}\right] = -64$
$\cos 4(x+30) = 0$

$4(x + 30) = \cos^{-1} 0 = 90 \quad x = -7.5^0$ or $360 - 7.5^0 = 352.5^0$; y $= 0$

Reading off preceding x intercepts or subtracting 45^0 in order to get preceding intercepts: 352.5, 0; 307.5, 0; 262.5, 0; 217.5, 0; 172.5, 0; 127.5, 0; 82.5, 0; 37.5,0

Example 4 Calculate the slope of the tangent passing through the points as stated on the cubic curve $y = x^3 - 9x^2 + 18x$. $x = 0$, $x = 3$, $x = 6$, $x = -3$

Solution: a) $\dfrac{dy}{dx} = 3x^2 - 18x + 18$ Substituting $x = 0$ into the differential $\dfrac{dy}{dx} = 18 =$ slope b) Substituting $x = 3$ into the differential $\dfrac{dy}{dx} = 3(3)^2 - 18(3) + 18 = -9$ c)$\dfrac{dy}{dx} = 3(6)^2 - 18(6) + 18 = 18$ d) $\dfrac{dy}{dx} = 3(-3)^2 - 18(-3) + 18 = 99$

As shown in Fig 6.5, the tangents passing the points are inclined at an angle with respect to the positive x-axis.

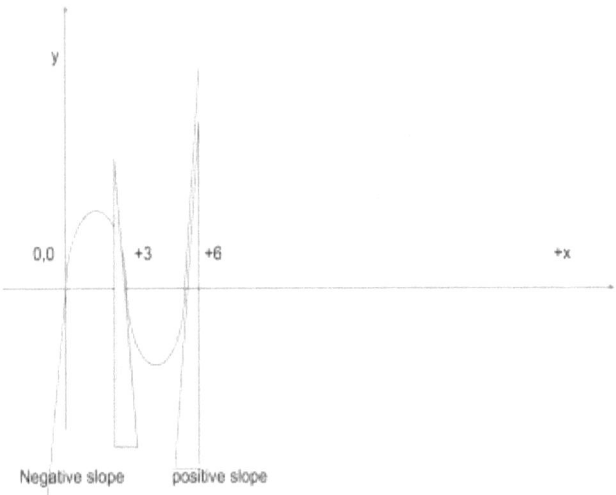

Fig 6.5 Graph of $y = x^3 - 9x^2 + 18x$ illustrating positive and negative slope.

Exercise 6.5

1. Apply the method of differentiation as follows: i) Calculate the maximum and minimum points. ii) Determine the points of inflexion. iii) Sketch and label the cubic curve a. $y = x^3 + 3x^2 - 16x - 48$ b. $y = 2x^3 + 7x^2 + 7x + 2$ c. $y = x^3 + 5x^2 - x - 5$ d. $y = x^3 - 2x^2 - 15x + 36$ e. $y = x^3 + 6x^2 + 11x + 6$

2. Calculate the gradient of the tangents passing through the specified points on each curve.

a. $y = x^3 + 3x^2 - 16x - 48$ at $x = -3$ b. $y = 2x^3 + 7x^2 + 7x + 2$ at $x = -2$ c. $y = x^3 + 5x^2 - x - 5$ at $x = -1$ d. $y = x^3 - 2x^2 - 15x + 36$ at $x = 1.5$ e. $y = x^3 + 6x^2 + 11x + 6$ at $x = -3$

6.10.2 Rates of Change Velocity, $\frac{ds}{dt}$, is the rate of change of distance travelled per unit time. Acceleration, $\frac{d^2s}{dt^2}$, is the rate of change velocity per unit time.

Differentiation is applied in calculating the velocities, v_A, v_B and v_C and acceleration, a_A, at the specified times t_A, t_B and t_C respectively as shown in the distance-time graph in Fig 6.6. The quadratic curve in Fig 6.6 is the equation of motion $s = ut + 0.5at^2$. Assuming an initial velocity of 3m/s and acceleration due gravity of -9.81m/s, the maximum vertical distance travelled, s_c, and velocities, v_A and v_B, are calculated thus.

The projectile path is given by the formula: $s = 3t - 4.905t^2$ the velocity at point A, $v_A = \frac{ds}{dt} = 3 - 9.81t$ the velocity at point B, $v_B = \frac{ds}{dt} = 3 - 9.81t$ the velocity at point C, $v_c = \frac{ds}{dt} = 3 - 9.81t$ Now, the last differential at the point C is equal to zero. This is because a tangent passing through the point C makes an angle of 0 degrees with respect to the x axis. $3 - 9.81t = 0$ $t = 0.3058$ seconds Substituting into $s = ut + 0.5at^2$ $s = (3m/s \times 0.3058 \text{ sec}) - 0.5(9.81m/s^2 \times 0.3058^2)$ $= 0.4587m$ The velocities at $t_A = 0.08955$ seconds and $t_B = 0.52207$ seconds are: $v_A = 3 - (9.81m/s \times 0.08955 \text{ seconds}) = 2.1215m/s$ $v_B = 3 - (9.81m/s \times 0.52207 \text{ seconds}) = -2.1215m/s$ $v_c = 3 - (9.81m/s \times 0.3058 \text{ seconds}) = 0m/s$ $a_A = \frac{d^2s}{dt^2} = -9.81m/s^2$ Rates of change calculations will be dealt with in detail in Engineering Mathematics N5 in the Applications of Differentiation section.

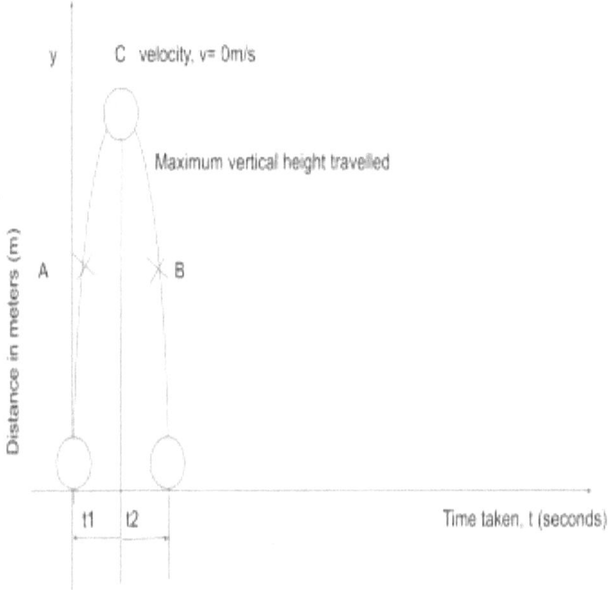

Fig 6.6 Projectile of a ball thrown into the air.

6.10.3 Limits

There is a relationship between the axis of symmetry, the range of a function and the limits of a function. See Fig 6.7 A typical question on limits is as follows. **Example 1** Calculate the limit of $y = x^2 + x - 6$ as x tends to -0.5. Mathematically, Lim $y = x^2 + x - 6 = -6.25$ x tends to -0.5

The range of the above function is $-6.25 \le y \le + \infty$. The axis of symmetry is: $x = - \dfrac{b}{2a} = -0.5$

Based on this explanation, substituting the value x = -0.5 into the equation $y = x^2 + x + 6$ will give us -6.25. $f(-0.5) = (-0.5)^2 + (-0.5) + 6 = -6.25$

Alternatively, another perspective of the above concept is based on substituting the value of y= -8 into the equation. Apply the quadratic formula after simplification. $-8 = x^2 + x + 6$ y= -8 is an undefined value and it does not intersect the quadratic function. The y value, y= -6.25, is therefore the limit of the equation $y = x^2 + x + 6$. This is minimum point or y coordinate of the minimum point.

The following example illustrates the relationship between domain, range and limits.

➤➤ **Example 2** Calculate a) $\lim y = \dfrac{8}{x}$. x 0 b) $\lim y = \dfrac{8}{x}$ x ∞

➤ Consider the table of values that have been used to plot the graphs is as shown below. Solution a) Based on the graph above (Fig 6.8), as we decrease the value of x gradually towards zero, our corresponding value of y tends to positive infinity. Consider the x coordinate: x = 0.0025 and substituting y = 8/0.0025 = 3200. This is true for the curve in the first quadrant. Likewise, substitution of x =

-0.0025 will give: y= 8/-0.0025 = -3200 $\lim y = \dfrac{8}{x} = \pm\infty$ x 0

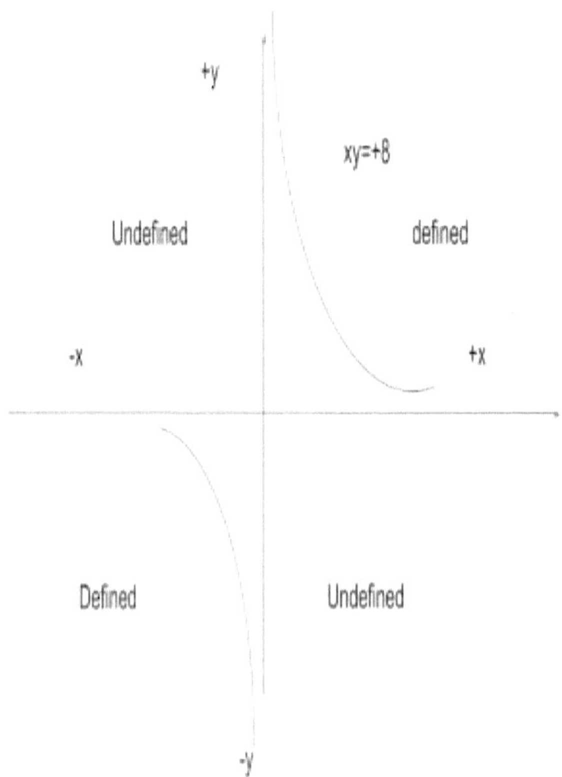

Fig 6.8 Graph of xy = 8

Table 6.8. 8 = xy

x	-50	-20	-15	-10	-5	0
y	-0.16	-0.4	-0.53	-0.8	-0.4	error
$dy/dx = -8x^{-2}$	-0.0032	-0.02	-0.036	-0.08	-0.32	

x	5	10	15	20	50
y	0.4	0.8	0.53	0.4	0.16
$dy/dx = -8x^{-2}$	-0.32	-0.08	-0.036	-0.02	-0.0032

b) $\text{Lim } y = \dfrac{8}{x} = 0 \ x \ \infty$

The value of differential for the minimum point of Example 1 is zero. This, therefore, is the limit of the curve based on the range $-6.25 \le y \le + \infty$. In the second example, the values of the derivatives tends to zero as x increases to infinity thus defining the asymptote $y = 0$. Fundamental limit problems should be solved based on the concept of the relationship between range and domain as discussed in example 1 and example 2 above.

Example 3 Calculate $\text{Lim } \dfrac{x^3-9x^2-26x-24}{x-4}$

$x \ 0 = \dfrac{(x-2)(x-3)(x-4)}{x-4} = (x-2)(x-3)$ substituting $x = 0 = (0-2)(0-3) = 6$

Example 4 Evaluate $\text{Lim } \dfrac{x^3-9x^2+18x}{x^2-3x}$

$x \ 6$

$= \dfrac{x(x-3)(x-6)}{x(x-3)} = x-6$ substituting $x = 6 = 6-6 = 0$ **Example 5** Find $\text{Lim } \dfrac{x+1}{x^3}$

$x \ 1 \ \dfrac{(x \div x^3)+(1 \div x^3)}{(x^3 \div x^3)} = \dfrac{x^{-2}+x^{-3}}{1} = x^{-2} + x^{-3}$ substituting $x = 1 = 2$

Example 5 may simply be evaluated by simply substituting the value of $x = 1$ into the given problem as shown below.

$$\frac{x+1}{x^3} = \frac{1+1}{1^3} = 2$$ When the algebraic fraction cannot be easily factorized and simplified, the denominator and the numerator should be divided by the algebraic term having the greatest power. In this case, the algebraic term is x^3.

Exercise 6.6 1. $\lim \dfrac{x^3-9x^2-26x-24}{x-4} \cdot x\,\infty$ 2. $\lim \dfrac{x^3-9x^2+18x}{x^2-3x} \cdot x\,1$ 3. $\lim \dfrac{x^3+x^2+x}{x^3} \cdot x\,\infty$ 4. $\lim \dfrac{x^3}{x^3+x^2+x} \cdot x\,0$ 5. $\lim e^{2x} \cdot x-1000$

6. $\lim e^{-3x} \cdot x\,\infty$ 7. $\lim \log x \cdot x\,\infty$ 8 $\lim \log x \cdot x\,0.0001$

Chapter 7 Integration

7.1 Introduction Integration is the opposite of differentiation. Alternatively, integration may be defined as summation of elemental strips.

One of the applications of integration is the calculation of plane areas whose boundaries are defined by a curve or straight lines.

Study the table of integrals below, Table 7.1.

7.2 Standard Formula for Integration Given an equation $y = ax^n$, the integral is $\int y \, dx = \frac{ax^{n+1}}{n+1}$.

The expression $\int y \, dx$ is the symbol for integral or integration.

Example 1 Find the integral of $y = 3x^3$. Find the values of a and n by comparing given equation with $y = ax^n$. a=3 and n = 3. Applying the formula for integration, we have: $\int y \, dx = \int 3x^3 dx = [3x^{3+1}]/[3+1] = 3x^4/4$ Students are advised to differentiate $3x^4/4$ in order to get our original cubic function.

7.3 Table of Integrals

Table 7.1

$y=$	$\int y\,dx$
ax^n	$[ax^{n+1}]/[n+1]$
Ka^x	$[ka^x]/\ln a$
$.e^x$	$.e^x$
$\ln x$	$x\ln x - x$
$\sin x$	$-\cos x$
$\cos x$	$\sin x$
$\tan x$	$-\ln \cos x$
$\cot x$	$\ln \sin x$
$\sec x$	$\ln(\sec x + \tan x)$
$\csc x$	$\ln(\csc x - \cot x)$

Example 2 Evaluate the integral of $y = \sin x$. $\int y\,dx = \int \sin x\,dx = -\cos x$

Example 3 Calculate $\int [x^3 + 2x^2 + 10x + 6]\,dx = \dfrac{x^4}{4} + \dfrac{2x^3}{3} + 5x^2 + 6x$

Example 4 Determine the integral of $\int (x+1)^4 dx = \int (x+1)^2(x+1)^2$

$dx = \int x^4 + 4x^3 + 6x^2 + 4x + 1\,dx = \dfrac{x^5}{5} + x^4 + 2x^3 + 2x^2 + x$

Example 5 Evaluate $\int \dfrac{x^2 - 9}{x+3}\,dx = \int \dfrac{(x+3)(x-3)}{x+3}\,dx = \int x - 3\,dx = \dfrac{x^2}{2} - 3x$

Example 6 Evaluate the integral of $y = \dfrac{\sin^2 x - 1}{\cos^2 x}$ $\quad \int \dfrac{x}{\cos^2 x}\,dx$

$= \int 1\,dx$

$= x$

Example 7 Find the integral of $y = \dfrac{\sin^2 x - 1}{(\sin x) + 1} = \int \dfrac{((\sin x) - 1)((\sin x) + 1)}{(\sin x) + 1}\, dx$

$= \int (\sin x) - 1\, dx$

$= -(\cos x) - x$ **Example 8** Calculate $\int y\, dx = \int (10X2^x)\, dx$ Compare with standard form from tables

$y = Ka^x = [10X2^x]/In2$

7.4 Integration by Substitution

The method of Integration by substitution is equivalent to the function of a function method in differentiation. This method of integration simplifies the integration by introducing a third variable u. As stated earlier in function of a function, students are required to write an integral in terms of u.

The following examples illustrate this concept.

Example 1 Find the integral of $y = \sin 3x$. $\int y\, dx = \int \sin 3x\, dx$. Let $u = 3x$ and $du = 3dx$ Write $\int y\, dx = \int \sin 3x\, dx$ in terms of u and du. $\dfrac{3}{3}\int \sin u\, dx = \dfrac{1}{3}\int \sin u\, (3dx)$

Careful consideration should be given to the term' $\dfrac{3}{3}$, which is equal to 1. This is necessary in order to obtain the term 3dx. The original problem does not have a coefficient of 3. Substitution: $du = 3dx$ since $\dfrac{du}{dx} = 3$ $\dfrac{1}{3}\int \sin u\, du = -\dfrac{1}{3}\cos 3x$

Example 2 Evaluate the integral of $y = 2\sin 5x\cos 5x$ Applying compound angles: $y = 2\sin 5x\cos 5x = \sin(5x + 5x)$ $\int y\, dx = \int \sin 10x\, dx$ Let $u = 10x$

$du = 10dx$

$$\frac{1}{10}\int \sin u\, du = -\frac{1}{10}\cos 10x$$

Example 3 Evaluate the integral of $y = \frac{9}{(x+3)^2}$ Let $u = x + 3$ and $du = dx$ $\int y dx = \int 9(x+3)^{-2} dx$ $\int y dx = 9\int u^{-2} du$ $-9\, u^{-1} = -9(x+3)^{-1}$

7.5 Applications of Integration

7.5.1 Constant of integration

Consider the following functions, whose differentials and integrals are as shown on the table below. Integrating the differential should give us our original expression in the first column.

Table 7.2 Constant of integration.

Function	$\dfrac{dy}{dx}$	$\int y dx$
$y = 3x + 5$	3	$3x + C$
$y = 3x^2 + 6x + 9$	$6x + 6$	$3x^2 + 6x + C$
$y = 3x^3 + 6x^2 + 10x + 15$	$9x^2 + 12x + 10$	$3x^3 + 6x^2 + 10x + C$

C is known as the constant of integration. This constant of integration is equal to the y intercept.

1. The y intercept x, y = 0,5 for the straight line $y = 3x + 5$

2. The y intercept x, y = 0,9 for the quadratic equation $y = 3x^2 + 6x + 9$

3. The y intercept x, y = 0, 15 for the cubic function $y = 3x^2 + 6x + 15$ Based on the above, the intercepts 5, 9 and 15, are the constants of intergration. The following examples will give a better understanding of the applications of integration.

Example 1 Find the equation of the straight line whose slope is 4 and passes through the point x,y = 0, 4. $\dfrac{dy}{dx} = 4$ dy= 4dx and $\int$dy

$= \int 4dx$ Integrating : $y = 4x + c$ Substituting $x, y = 0, 4$ into equation we have: $4 = 4(0) + C.$ $C = 4$ $y = 4x + 4$

Example 2 Calculate the equation of the curve given that: i) $\frac{dy}{dx} = 6x + 7$ ii) the curve passes through the point $(x, y) = (0, 2)$. $\int dy = \int 6x + 7 \, dx$ and $y = 3x^2 + 7x + C$ Substituting the $(x,y) = (0,2)$ into the equation to find C. $2 = 3(0)^2 + 7(0) + C$ $C = 2$ Therefore, the equation $y = 3x^2 + 7x + 2$

Example 3 A cubic equation is given by the formula $y = ax^3 + bx^2 + cx + d$. Find the equation of this function given that : i) $\frac{dy}{dx} = 3x^2 - 18x + 26$ ii) y intercept $= -24$ Solution:i) $\int dy = \int 3x^2 - 18x + 26 \, dx$ $y = x^3 - 9x^2 + 26x + C$ Substitute $x, y = 0.$ -24 $C = -24$ and $y = x^3 - 9x^2 + 26x - 24$ Refer to Fig 6.4.

7.5.2 Calculation of Areas

The Limits of Integration The limits of integration are also known as definite integrals. This refers to the domain that applies to a specific area that has to be calculated. The Limit of Integration may also be applied to calculate an area by considering the limits of the range that applies to the area in question.

Example 1 Calculate the area enclosed by the x axes and the two lines in Fig 7.1. by considering limits of integration with respect to: i) domain ii) range.

Solution: Integrating with respect to x means that the limits are defined by the x values and the elemental strip is perpendicular to the x axis. Integrating with respect to y refers to the fact that the elemental strip is perpendicular to the y axis. As a result of this, the definite integral is defined by the y values. Points of intersection of both lines is evaluated by: $6x + 12 = 12 - x$ Solving for x : $x = 0$ and $y = 12$ For $y = 12 - x$: substitute $y = 0$, $x = +12$ For $y = 6x + 12$: substitute $y = 0$, $x = -2$ i) Area $= \int y dx$ Total Area $=$ Area AOB $+$ Area AOC

$$= \int_{-2}^{0} 6x + 12 \, dx + \int_{0}^{12} -x + 12 dx = (3x^2 + 12x)_{-2}^{0} + \left(-\frac{x^2}{2} + 12x \right)_{0}^{12}$$

$$= 12 + 72 = 84 \text{ units}^2$$

ii) Based on the two equations $y = 12-x$ and $y = 6x + 12$ make x the subject of the formula. $x = 12-y$ and $x = \frac{y}{6} - 2$. We can now calculate the area by applying the y points of intersection on the integral. $\int_{0}^{12} 12 - y \, dy + \int_{0}^{12} \frac{y}{6} - 2 \, dy$

$$= -\left[12y - \frac{y^2}{2} \right]_{0}^{12} + \left[\frac{y^2}{12} - 2y \right]_{0}^{12}$$

$$= 84 \text{ units}^2$$

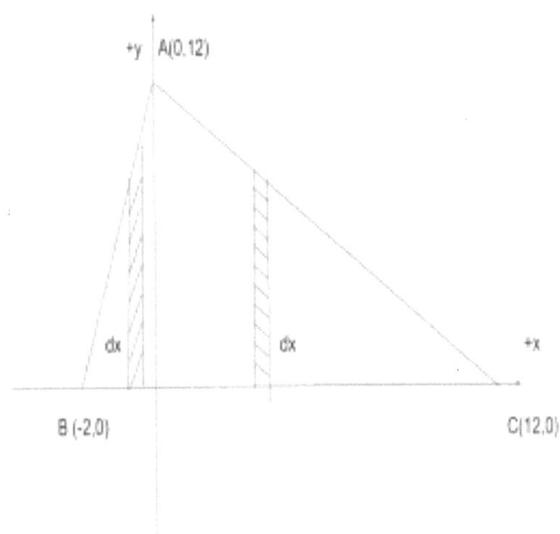

Fig 7.1 Integrating with respect to dx.

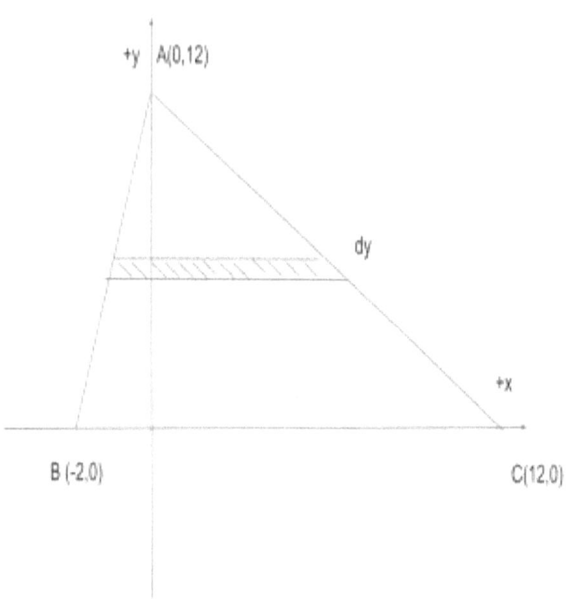

Fig 7.2 Integrating with respect to dy.

The area of AOB is negative because the base is -2 units. The negative sign ensures that we do not subtract Area AOB from Area AOC.

Example 2 A quadratic curve has the formula $y = x^2 - 10x - 24$. Evaluate the area enclosed by the quadratic curve and the x axes by: a) Calculating the turning points b) Evaluating the x intercepts c) Sketching the curve by labeling the following i) the representative element dx. ii) the turning points iii) the y intercept. Solution: a) Turning points: $\frac{dy}{dx} = 2x - 10 = 0$ $y = f(5) = (5)^2 - 10(5) - 24 = -49$ Substitute x into original equation b) x intercepts: factorizing $y = (x -12)(x +2)$ when y=0, x = 12 and x = -2 c) substitute x = 0 into the original equation y = -24.

The formula for calculating area $\int y\,dx$ is based on the area of the representative strip or element as shown in the diagram above. (Area =length X breadth) i) $\int y\,dx = \int^{12}_{-2} x^2 - 10x - 24\ dx = \left[\dfrac{x^3}{3} - 5x^2 - 24x\right]^{12}_{-2} = \left[\dfrac{12^3}{3} - 5(12)^2 - 24(12)\right] - \left[\dfrac{(-2)^3}{3} - 5(-2)^2 - 24(-2)\right] = -432 - 25.33 = -457.33\ \text{units}^2$ -

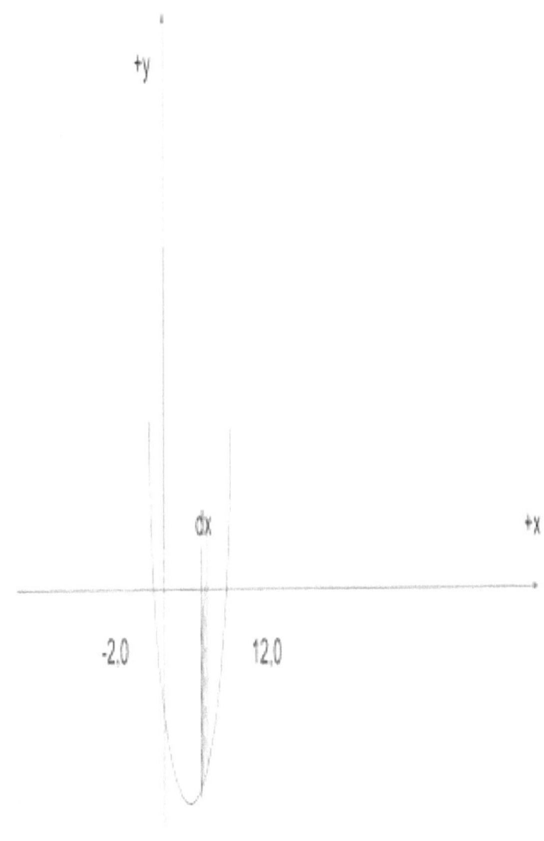

Fig 7.3 Graph of $y = x^2 - 10x - 24$

Alternatively, the area may be calculated in two steps. $\int_{-2}^{0} x^2 - 10x - 24\, dx = [\frac{x^3}{3} - 5x^2 - 24x]_{-2}^{0} = 0 - [\frac{(-2)^3}{3} - 5(-2)^2 - 24(-2)]$

$= -25.33$ units $\int_{0}^{12} x^2 - 10x - 24\, dx = [\frac{x^3}{3} - 5x^2 - 24x]_{0}^{12}$

$= -432$ units Total area $= -25.33 - 432$

$= -457.33$ units^2 Apply this two-step method to the previous example.

Example 3 Calculate the area enclosed by $y = \cos 2x$ and the x axes for the domain $0 \le x 180$. Solution Sketch the graph by completing a table of values x and y (x intervals 30^0). Based on the graph below, the area can be calculated thus: $\int_{0}^{45} \cos 2x\, dx$ Applying integration by substitution Let $u = 2x$ and $du = 2dx$

$\frac{1}{2}\int_{0}^{45} \cos u\, (2dx) = \frac{1}{2}\int_{0}^{45} \cos u\, du = \frac{1}{2}[\sin 2x]_{0}^{45} = 0.5$ units^2

$\int_{45}^{135} \cos 2x\, dx = \frac{1}{2}[\sin 2x]_{45}^{135} = \frac{1}{2}[[\sin 2 \text{X} 135] - [\sin 2\text{X}45]]$

$= \frac{1}{2}[-1 -1] = -1$ unit^2 $\int_{45}^{180} \cos 2x\, dx = \frac{1}{2}[\sin 2x]_{135}^{180} = \frac{1}{2}[[\sin 2\text{X}180] - [\sin 2\text{X}135]] = 0.5$ units^2

Total Area $= 0.5 + 1 + 0.5 = 2$ units^2 Please note that the -1 unit^2 is now $+1$ unit^2. The total area considering the original positive and negative signs is zero units^2 which is not applicable.

Table 7.3 Table of values $y = \cos 2x$.

x	0	15	30	45	60	75	90
Y=cos2x	1	0.866	0.5	0	-0.5	-0.866	-1

x	105	120	135	150	165	180
Y=cos2x	-0.866	-0.5	0	0.5	0.866	1

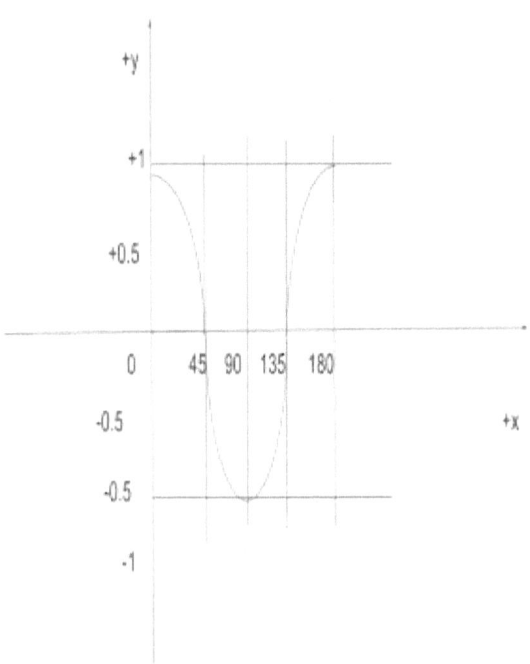

Fig 7.4 Graph of y = cos 2x

Exercises 7.1

1. Evaluate the following integrals: a) $\int \csc x \cot x \, dx$ b) $\int \sqrt[56]{x}$
    dx c) $\int 3x^7 \, dx$ d) $\int 10x^3 + 16x^2 + 2x + 15 \, dx$ e) $\int (\cos^2 16x -$
    $\sin^2 16x) dx$ f) $\int (\cos^2 x + \sin^2 x) \, dx$ g) $\int 100 \cos 34x \, dx$ h) $\int$
    $3^{15x} \, dx$ i) $\int 2\sin 16x \cos 16x \, dx$ j) $\int e^{15x} \, dx$ k) $\int 17\text{-}3x \, dx$ l) $\int$
    $\dfrac{x^2 - 100}{x + 10} \, dx$

2. Calculate the areas enclosed as stated by method of

integration. Sketch and label the x and y intercepts, the turning points, the enclosed area and the elemental strip. a) the lines $y = x - 20$, $y = -3x + 5$ and the x axes. b) the lines $y = x$, $x = 6$ and the x axes. c) the lines $y = x - 20$, $y = -3x + 5$ and the y axes d) the lines $y = x + 1$, $y = 6x - 10$ and the y axes. e) the curve $y = x^2 + 7x + 10$ and the x axes f) the curve $y = 2x^2 + 3x - 9$ and the x axes. g) the curve $y = 6x^2 + x - 1$ and the x axes. h) the curve $y = e^{6x}$, $x = 3$, $x = 6$ and the x axes. i) the curve $y = 6e^{6x}$, $x = 2$, $x = 5$ and the x axes. j) $y = \sin 2x$ and the x axes for the domain $0 \leq x \leq 180$. k) $y = 2\cos 3x$ and the x axes for the domain $0 \leq x \leq 120$. l) $y = x^3 - 9x^2 + 26x - 24$ and the x axis.

m) $y = x^3 + 3x^2 - 16x - 48$ and the x axis. n) $y = 2x^3 + 7x^2 + 7x + 2$ and the x axis. o) $y = x^3 + 5x^2 - x - 5$ and the x axis.

Answers

Ex 1.1

1. a)

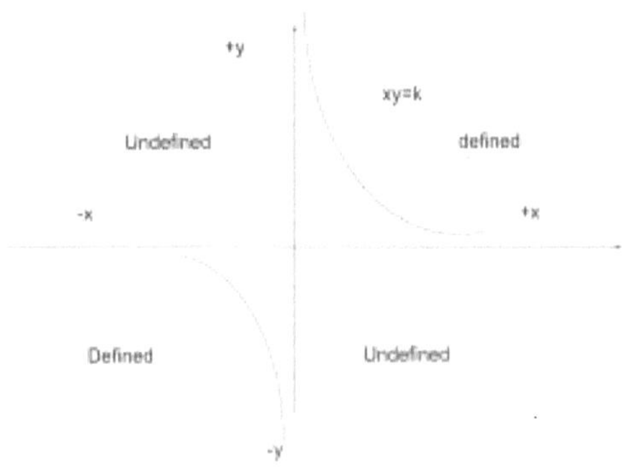

1. b)

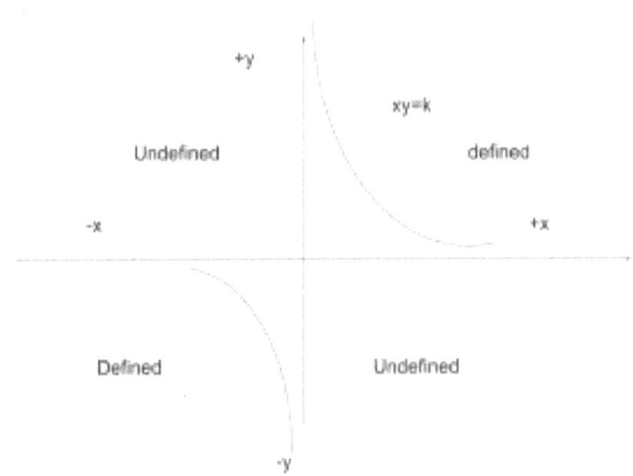

1. c)

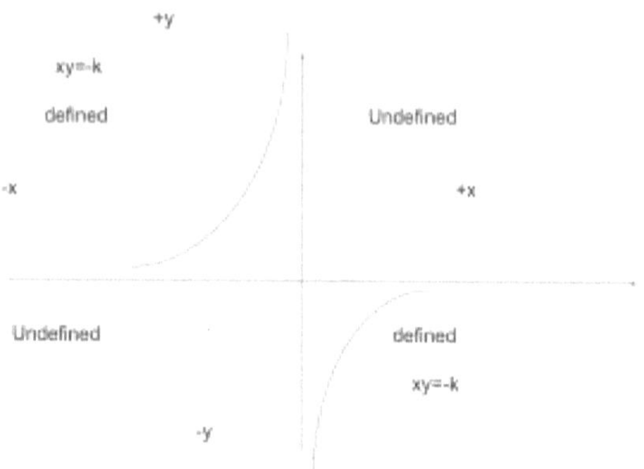

1. d)

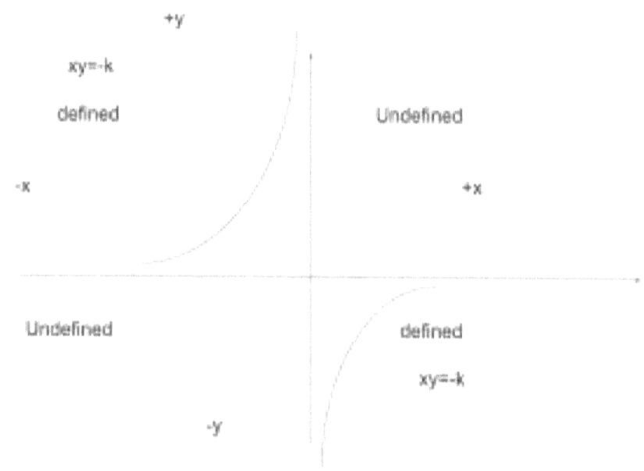

2.a) y = 2x + 5 b) y= -x +13 c) y = 0.5x + 0.5 d) y= x – 5 e) y= 0.364x – 3.46 f) y = 5.67x – 16 4 a) y= 6x – 14 b) y = 1.417x – 4.167 c) y= 0.9x – 0.3 d) y= - 5x + 15 e) y= x – 1 f) y= 1.143x + 13.857 5. a) $-\frac{1}{6}$;170.54^0 b) 6.857; 81.703^0 c) 5 ; 78.69^0 d) $\frac{2}{15}$; 7.59^0 e) 10 ; 84.29^0 f) -6 ; 99.46^0

| Y=5.67x - | Y=-x | Y=2x +5 | Y= x -5 |

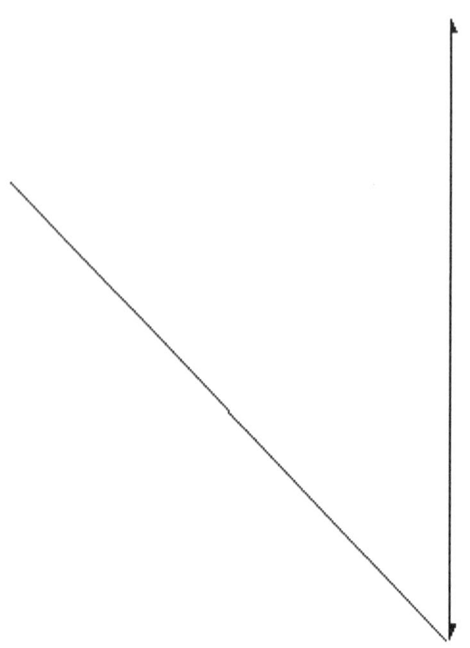
Y=0.5x +0.5
y=0.364x-
3.46

	X_1, y_1	X_2, y_2	Base Horizontal Distance	Height Vertical Distance	Slope
$y = 2x + 5$	-2.5, 0	0 , 5	0 – -2.5 = 2.5	5 – 0 = 5	2
$y = -x + 13$	13, 0	0 , 13	0 – 13 = -13	13 - 0 = 13	-1
$y = 0.5x + 0.5$	-1 , 0	0, 0.5	0 - -1 = 1	0.5 – 0 = 0.5	0.5
$y = x - 5$	0 , -5	5 , 0	5 – 0 = 5	0 - -5 = +5	1
$y = 0.364x - 3.46$	0, -3.46	9.5 , 0	9.5 – 0 = 9.5	0 - -3.46 = 3.46	0.364
$y = 5.67x -16$	0, -16	3 , 1	3 – 0 = 3	1 - -16 = 17	5.67

6. a) $y = \frac{1}{2}(x + 1)$ b) $y = -\frac{3}{4}x - \frac{5}{2}$ c) $y = -\frac{1}{3}(x + 1)$ d) $y = -0.5x$ e) $y = -0.25x$ f) $y = \frac{1}{3}(x + 10)$ 7. a) $y = 0.167x + 8$ b) $y = -0.146x + 7.876$ c) $y = 5x -23$ d) $y = 0.133x + 6.2$ e) $y = 10x – 53$ f) $y = -6x + 43$ 8. a) $y = 6x -13$ b) $y = 6.85x - 14.69$ c) $y = -0.2x - 0.6$ d) $y = -7.5x + 14$ e) $y = 0.1x - 0.8$ f) $y = 0.167x - 1.333$ 9. a) $y = -0.167x – 3.33$ b) $y = -0.146x - 3.292$ c) $y = 5x + 7$ d) $y = 0.133x - 2.73$ e) $y = 10x + 17$ f) $y = -6x - 15$ 10. a) $y = 6x + 8$ b) $y = 6.85x + 8.85$ c) $y = -0.2x + 1.8$ d) $y = -7.5x - 5.5$ e) $y = -0.1x + 1.9$ f) $y = \frac{1}{6}(x + 13)$ Ex 1.2 1. a) $r = 0.7071$; $-0.7071 \leq x \leq 0.7071$ b) $r = 100$; $-100 \leq x \leq 100$ c) $r = 60$; $-60 \leq x \leq 60$ d) $r = 0.866$; $-0.866 \leq x \leq 0.866$ e) $r = 1.5$; $-1.5 \leq x \leq 1.5$ f) $r = 0.926$; $-0.926 \leq x \leq 0.926$

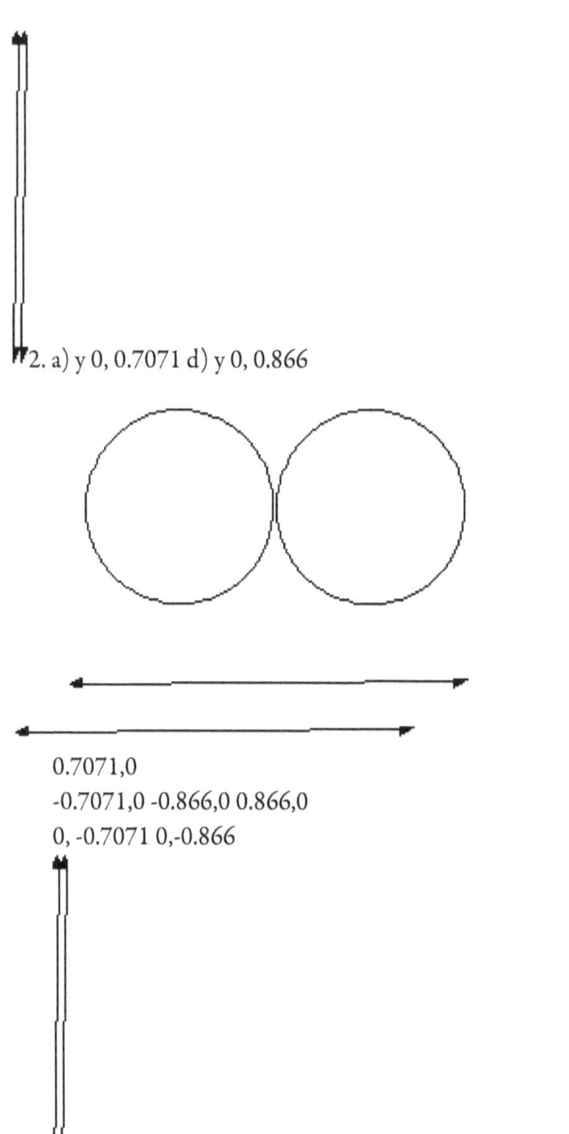

2. a) y 0, 0.7071 d) y 0, 0.866

0.7071,0
-0.7071,0 -0.866,0 0.866,0
0, -0.7071 0,-0.866

b) y 0,100 e) y 0, 1.5

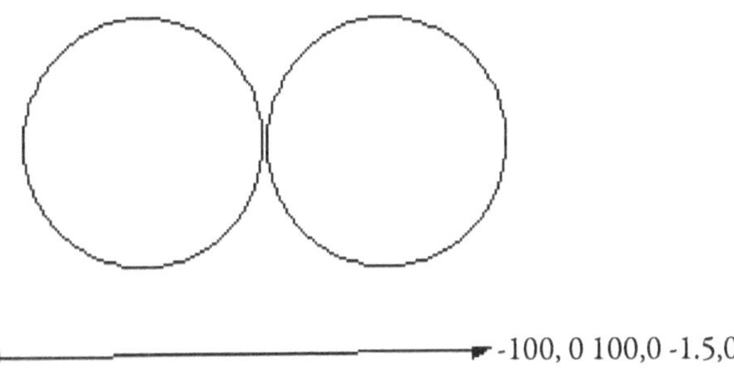

Xx 0, -100 0, -1.5

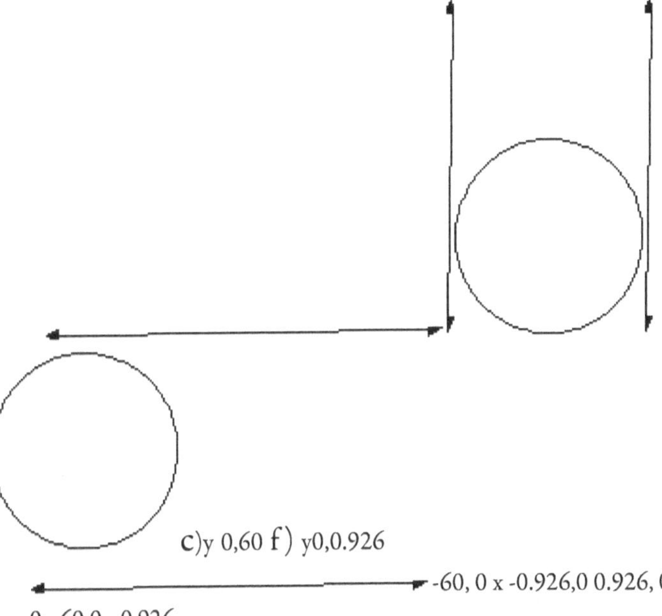

0, -60 0, -0.926

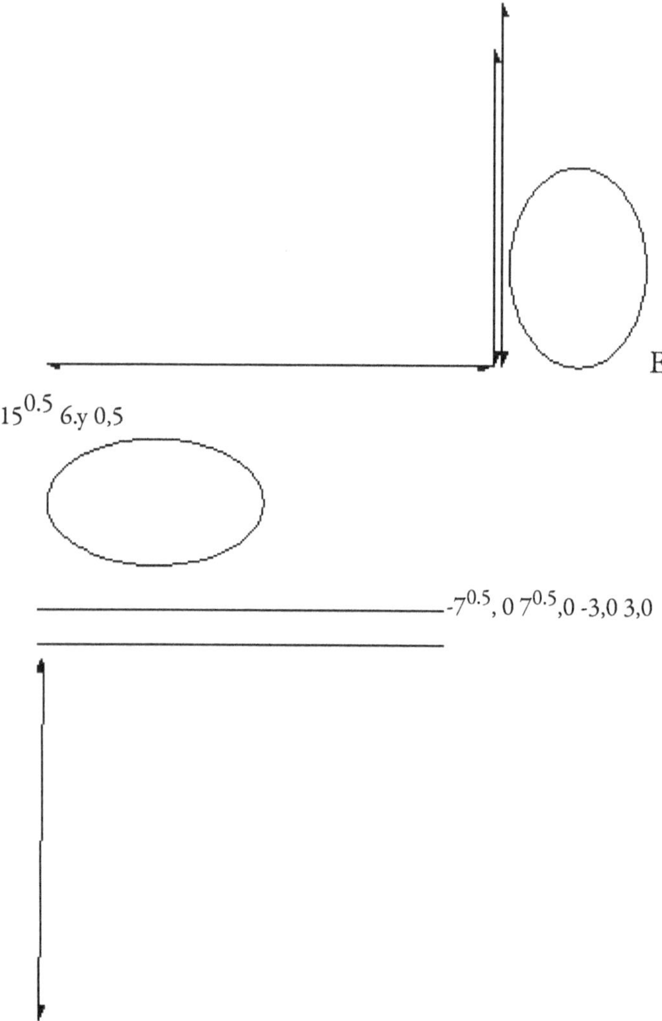

2. y 0, $24^{0.5}$7. Y 0,0.111

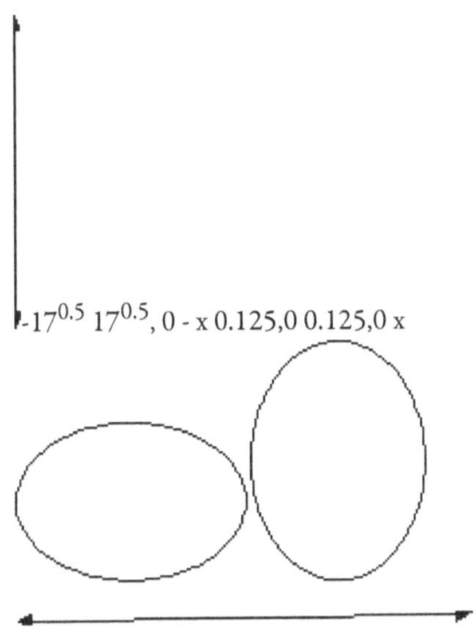

0, -24^0.5·0,- 0.111

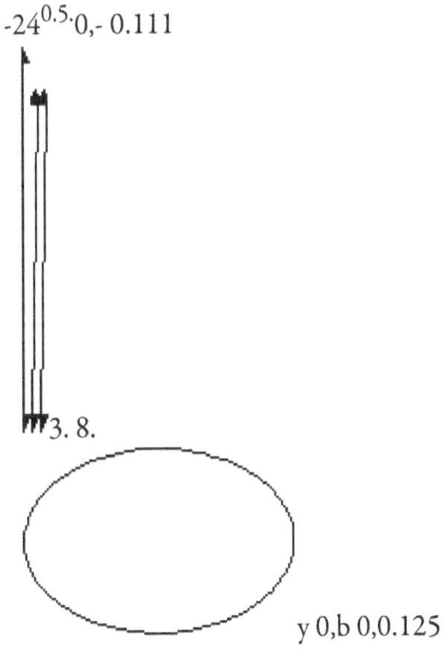

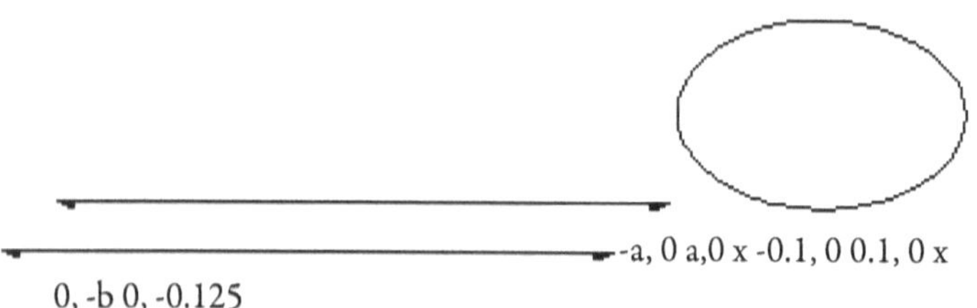
-a, 0 a,0 x -0.1, 0 0.1, 0 x
0, -b 0, -0.125

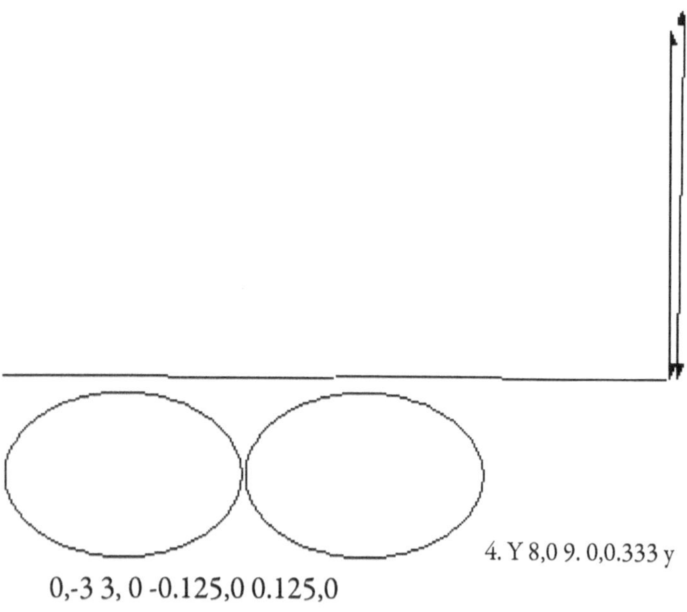
4. Y 8,0 9. 0,0.333 y
0,-3 3, 0 -0.125,0 0.125,0
-8,0 0, -0.333

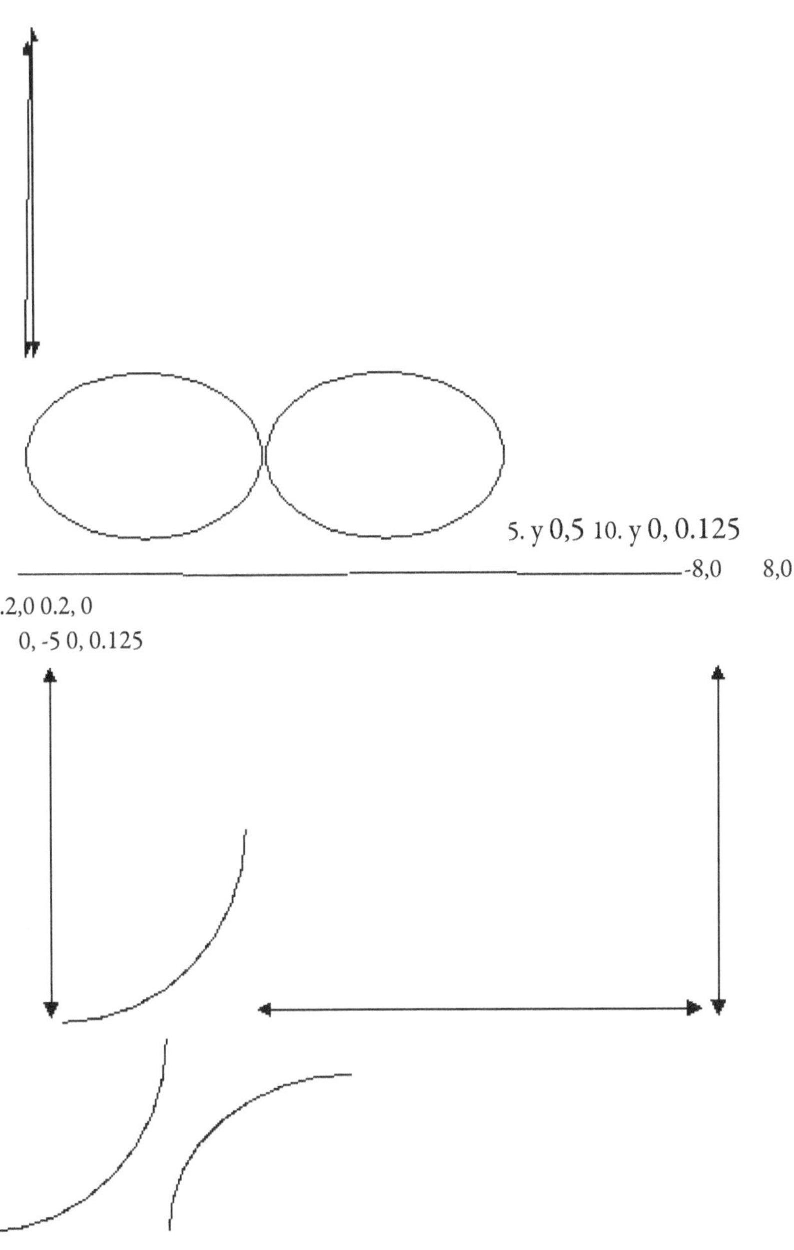
5. y 0,5 10. y 0, 0.125
-8,0   8,0
-0.2,0 0.2, 0
0, -5 0, 0.125
Ex 1.4

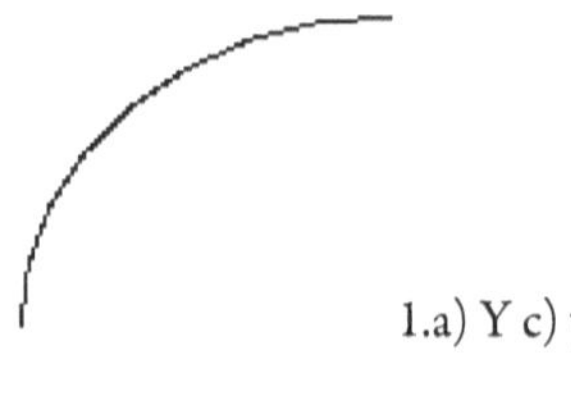

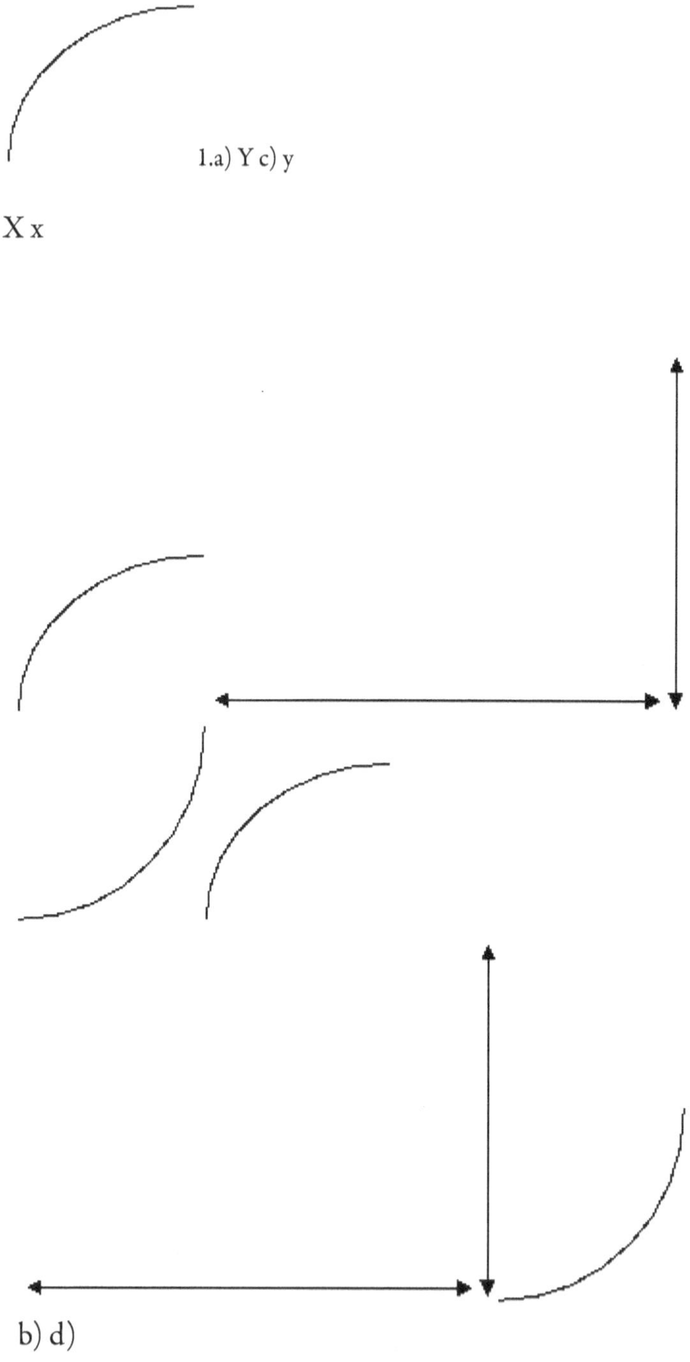

1.a) Y c) y

X x

b) d)

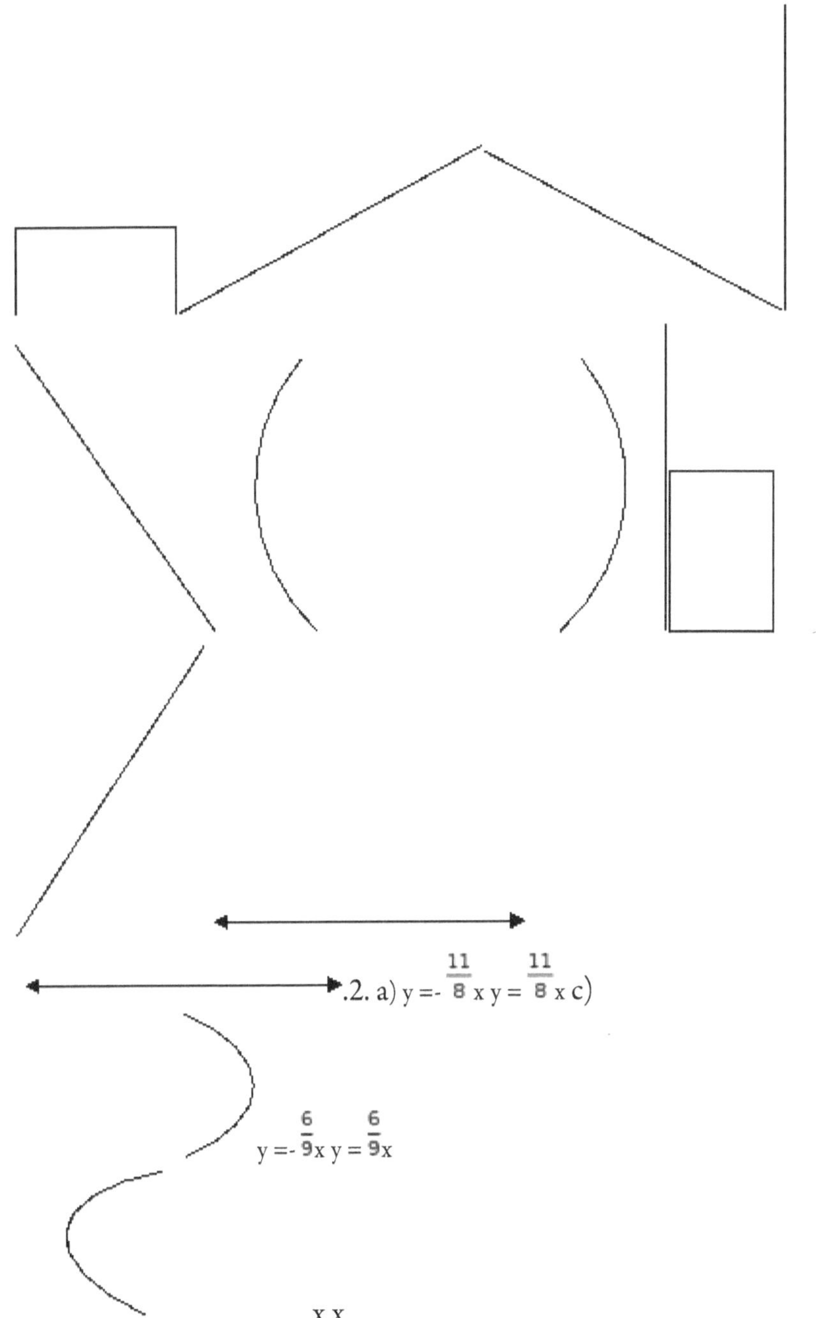

.2. a) $y = -\dfrac{11}{8}x$ $y = \dfrac{11}{8}x$ c)

$y = -\dfrac{6}{9}x$ $y = \dfrac{6}{9}x$

x x

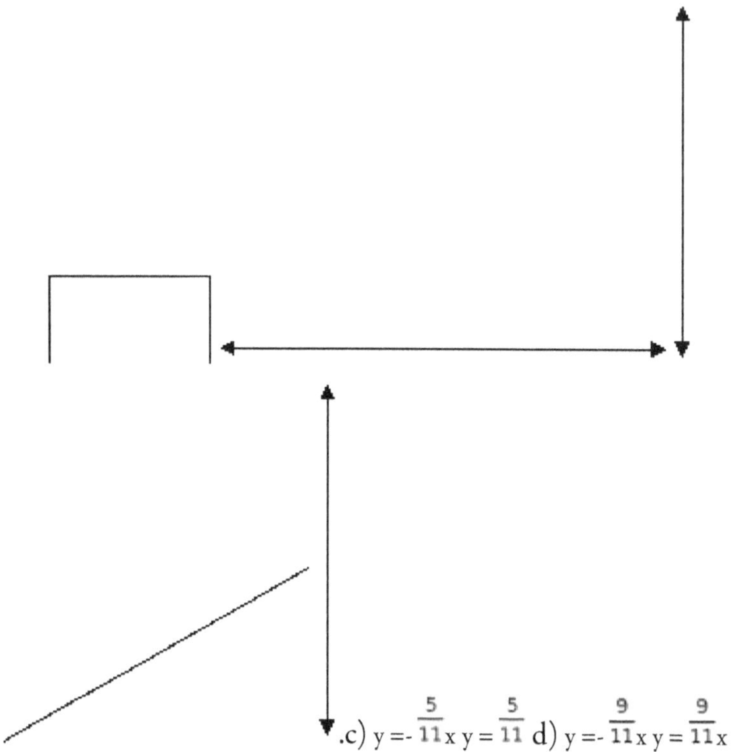

.c) $y = -\dfrac{5}{11}x \; y = \dfrac{5}{11}$ d) $y = -\dfrac{9}{11}x \; y = \dfrac{9}{11}x$

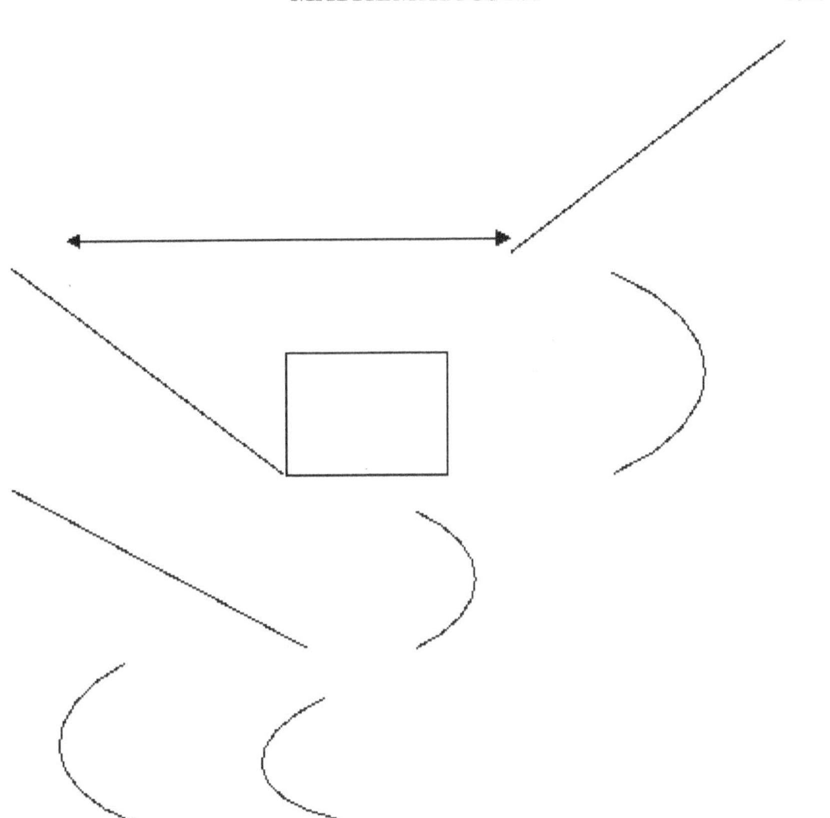

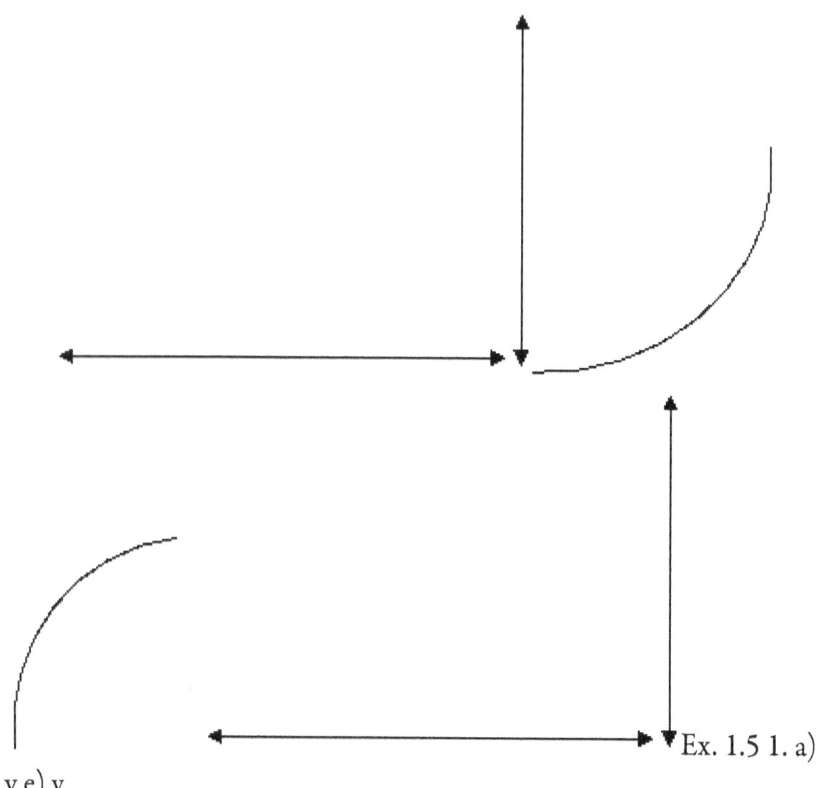
Ex. 1.5 1. a)
y e) y

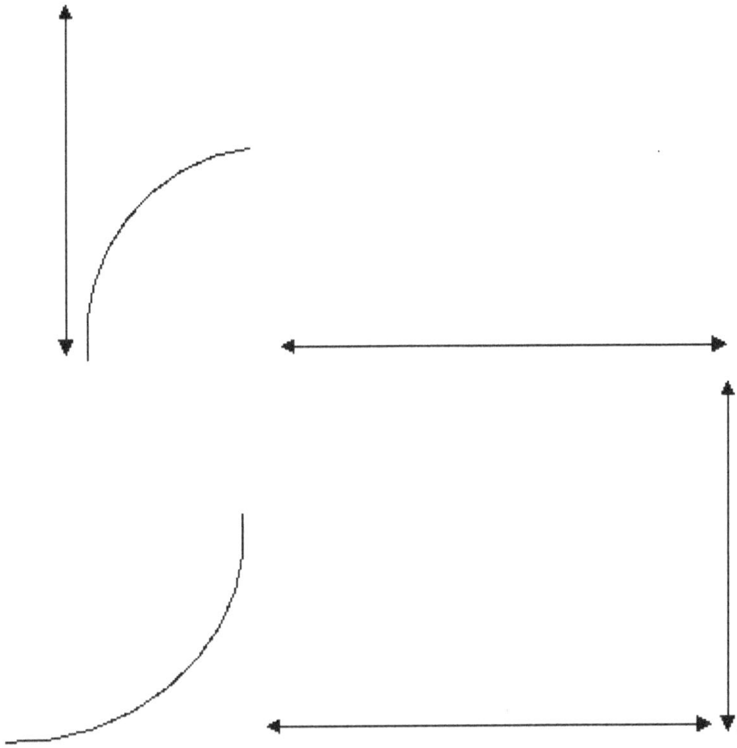

1, 0 x

.b) y f) y
0,2 x 0.5, 0 x

0,1   x

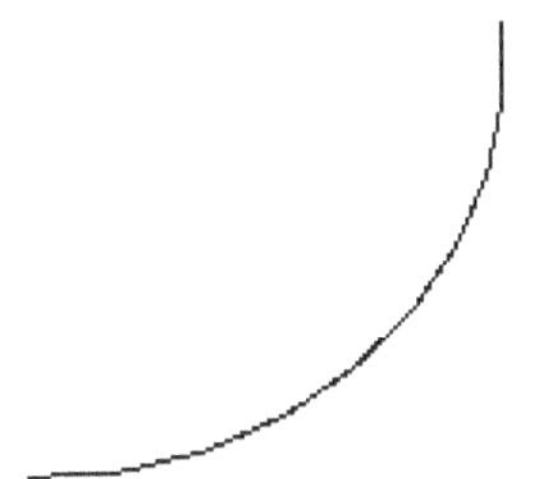

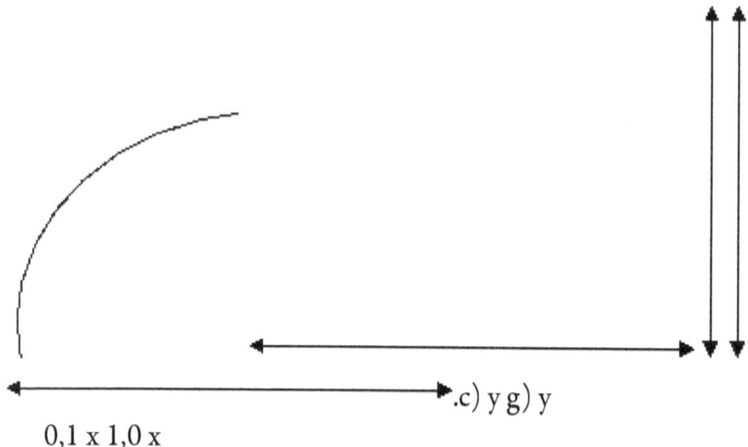

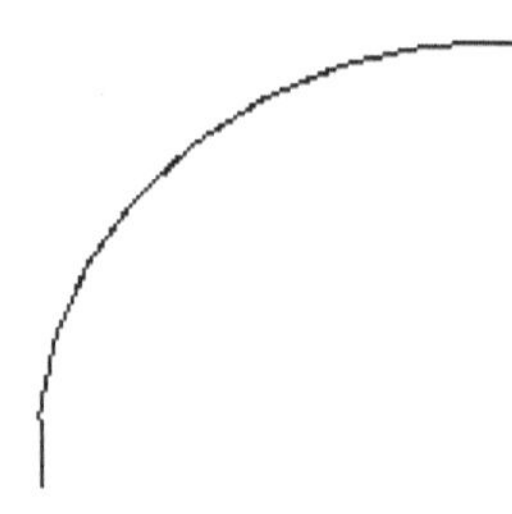

0,1 x 1,0 x

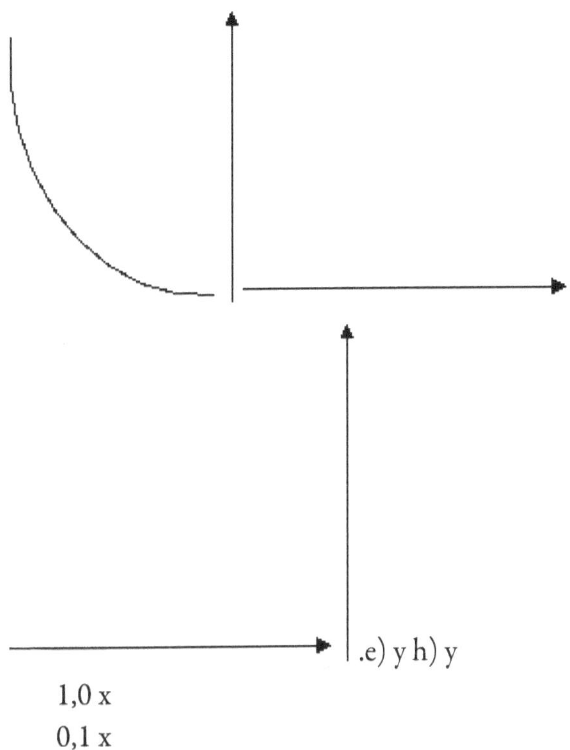

Ex 1.6

1.a) 315

b 33.69 , 213.69 c) 33.75; 123.75 2. a) 120 ; 4 ; 3; 45^0 leading ; 180^0 ; 4 ; 2 ; $0 \leq x \leq 360$ b) 360 ; 3 ; 1 ; 360; 2; 1; 0 phase shift c) 180 ; 4 ; 2 ; 45^0 leading ; $0 \leq x \leq 360$

Ex 1.7 1. x =-3 ; 4 ; -4; -3.517 ; 1.517 ; y = -48 ; 1.877 ; -61.877 2. x= -0.5 ; -2 ; -1 ; -1.608 ; -0.726 ; y = 2 ; 0.528 ; - 0.158 3. x= 1 ; -1 ; -5 ; -3.43 ; 0.097 ; y = -5 ; 16.9 ; -5.05 4. x= 3 ; 3 ; -4 ; -1.667 ; 3 ; y = 36 ; 0 ; 50.81 5. x= -1 ; -2 ; -3 ; -2.577 ; -1.423 ; y = 6 ; 0.385 ; - 0.385

Ex 2.1 1. x= 3.267 ; y= 5.199 2. X= -1.789 ; y =1.263 3. X= 1.796 ; y= 0.2592 4. x= -10 ; y= -20 5. x= -1.5258 ; y= 4.0898 ; z= -1.4872

6. x= 6.7683 ; y= 2.5104 ; z= -3.2317 7. x= -0.0068 ; y=0.6138 ; z = 0.3276 8. x= 2.8648 ; y= -1.8919 ; z=-10.1889 9. x= 2.333 ; - 0.1429 10. x = 3.3 ; 1.5 11. X= 1 ; -3.6 12. X= 2 ; -1.8 13. X= - 0.333 ; - 2.5 14. X= 2.333 ; - 0.1429 15. X = 3.3 ; 1.5 16. X = 3.3 ; 1.5 17. X= 1 ; -3.6 18. X= 2 ; -1.8 19. X= 2.333 ; - 0.1429 20. X= 3.3 ; 1.5 21. X = 1 ; -3.6 22. X = 2 ; -1.8 23. X= -0.333 ; -2.5 24. R11.20 ; R12.50 25. 79.64km/h ; 104.64km/h; 92.14km/h26. 22.5 km/h ; 15 km/h 27. 184km/h ; 200km/h 28. 4m ; 16m 29. 79.639 km/hr 30. 64 ; 66 32. 12cm ; 16cm ; 20cm Ex 2.2 1. $2^{0.8}$ 2. 9 3. 11^4 4. $3^{1.3}$ 5. 27 6. 8 7. -0.3611 8. 1.3 9. -2.667 10. -40.21 11. 1.51 12. 3.44 13. 8.55 14. 3.16 15. 6.25 16. 116.13 17. 1.059 18. 2.49

Ex 2.3 1. $C = \{ \frac{E}{2\pi f} \}^2 \frac{1}{L}$ 2. $\phi = 2\pi [\frac{R}{2L}]^2$ 3. $R = 2L \sqrt{\frac{1}{LC} - [2\pi f]^2}$ 4. $n = \frac{\log[\frac{S(r-1) + a}{a}]}{\log r}$ 5. $\mu = \frac{\ln[\frac{T1}{T2}]}{\theta}$ 6. $X = 8^y$ 7. $h = v^2/2g$ 8. $n = \frac{\log \frac{A}{P}}{\log (1+r)}$

Ex 3.1 1. X =17 ; y= 7 2. X= -28 ; y= -22 3. X= -3.5 ; y = 1.25 4. X= -7.67 ; y = -12.005 5. X= 0.818 ; y=1.091 6. X= 1.4474; y=-0.210527. X= 2; y=7 8. X=1; y=4 Ex 3.2 1. X= -1.5258 ; y= 4.0898 ; z= -1.4872 2. X= 6.7683 ; y= 2.5104 ; z= -3.2317 3. X= -1.22068 ; y=0.6416 ; z = -0.01724 4. X= 2.8648 ; y= -1.8919 ; z=-10.1889

Ex 4.1 1. i) 0.948 ii) 0.695 iii) 0.658 2. i) 0.966 ii) 0.793 iii) 0.924 3. i) -0.866 ii) 0.866 iii) -1.732 4. 0.5, -0.5 ; - 0.866, -0.866 ; -1.732 , 1.732 ; -0.577, 0.577 5. i) 1.1547, -1.55 ii) 2, 2 iii) 0.577 , - 0.577 6. i) -1.155, - 1.155 ii) 2, 2 iii) 0.577, -0.577 7. i) 0.3194, 2.965 ; 0.3196, -2.965 ii) -0.719, -0.967 ; -0.719, 0.967 iii) 0.753 , 0.873 ; 0.753, 0.873

8. $\dfrac{(2-\sqrt{2})^{0.5}}{2}$; $\dfrac{-(2+\sqrt{2})^{0.5}}{(2-\sqrt{2})^{0.5}}$ 9. $\dfrac{-(2-\sqrt{2})^{0.5}}{2}$; $\dfrac{-(2-\sqrt{\sqrt{2}})^{0.5}}{(2+\sqrt{2})^{0.5}}$ 10. $\dfrac{(2+\sqrt{3})^{0.5}}{2}$;

$\dfrac{-(2-\sqrt{3})^{0.5}}{(2+\sqrt{3})^{0.5}}$ 11. -2.83 ; 2.83 12. 1.8073; -1.8073

Ex 4.2 1. 35.54 , 215.54 2. 35, 215 3. 199.45, 340.55 4. 90 5. 186.08, 353.92 6. 11.31, 191.31; 168.69, 348.69 7. 8. 41.6, 221.6 ; 141.75, 321.75 9. 30, 150 ; 210, 330 10. 221.39, 318.613 11. 18.42, 198.42 ; 161.58, 341.58 12. 90, 270, 450, 630 13. x = 10 , y = -10 14. X= 45 ; y = 0 15. X= 67.5 , y= 112.5 16. X= 105 ; 165 ; y = -105, -165 17. Find LCM and simplify 18. $2\tan\Theta\sin^2\Theta$ 19. $(2-\sin^2\Theta)\cot^2\Theta$

Ex 5.1

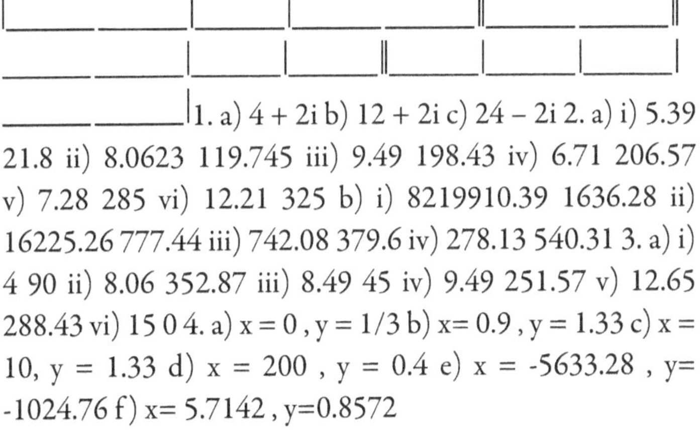

1. a) 4 + 2i b) 12 + 2i c) 24 – 2i 2. a) i) 5.39 21.8 ii) 8.0623 119.745 iii) 9.49 198.43 iv) 6.71 206.57 v) 7.28 285 vi) 12.21 325 b) i) 8219910.39 1636.28 ii) 16225.26 777.44 iii) 742.08 379.6 iv) 278.13 540.31 3. a) i) 4 90 ii) 8.06 352.87 iii) 8.49 45 iv) 9.49 251.57 v) 12.65 288.43 vi) 15 0 4. a) x = 0 , y = 1/3 b) x= 0.9 , y = 1.33 c) x = 10, y = 1.33 d) x = 200 , y = 0.4 e) x = -5633.28 , y= -1024.76 f) x= 5.7142 , y=0.8572

Ex 5.2

1. 478 317.957 N 2. a) 14.359 – 11.903i Ω b) 10.723 -320.34 A

————‖‖————|————3. M = 0 + 0i ; 0 0 4. W= 2 + 2i ;
2.828 45

————‖————————N = 1.131 + 1.131i ; 1.6 45 X =
5.182 + 5.182i ; 7.328 45

————|————|————|O= -2.051 + 4.313 ; 4.776
115.433 Y 4.051 + 6.313i ; 7.501 57.312

————|————|————|P = -3.182, 3.182 ; 4.5 135 Z =
0.869 + 3.131i; 3.249 74.49

5. 251.314 km/hr ; 29.497 ° NE

Ex 6.1 1. a) $3x^2 + 6x - 16$ b) $9x^2 + 14x$ c) $3x^2 + 10x -1$ 2. a) $\sec^2 x$ b) $- \csc^2 x$ c) $- \csc x \cot x$ d) $\sec x \tan x$ e) $-(6x - 3)(2x^3 - 3x)^{-2}$ f) $9(3x + 10)^2$ 3. a) $42\cos 7x$ b) $10e^x x^3(x + 4)$ c) $2x^2(3\cos x + x\sin x)(\cos x)^{-1}$ d) $-6\sin x \cos^5 x$ e) $21(x^3 + 3x^2 -16x -48)^{20}(3x^2 + 6x -16)$ f) $80e^{8x}$

Ex. 6.2 1. $4\sin^3 x \cos x$ 2. $- \csc^2 A$ 3. $3\tan^2 A \sec^2 A$ 4. $2\cos x \sin (x - 3)$ 5. $2\cot x \csc^2 x$ 6. $10\csc^2\theta(\cot^2 \theta - 2\csc^2 \theta\csc^2 2\theta -1)$ 7. $-2\cos A \sin A$ 8. $2\sin A\cos A$ 9. $-45 \csc^2 3\theta$ 10. $-8\csc^8 A \cot A$ 11. $-\csc^2 x$ 12. $-4\cot^3 x\csc^2 x$ Ex. 6.3 1. $\dfrac{3}{x\ln 10}$ 4. $75e^{3x} + \dfrac{25}{x}$ 2. $120e^{2x}$ 5. $\dfrac{10}{x}$ 3. $\dfrac{20}{x}$ 6. $8e^{4x} - \dfrac{8}{7x^{10}\ln 10}$ Ex. 6.4 1. $(2x)^{0.25} + \dfrac{(2x)^{-0.75}}{4} - \dfrac{3(2x)^{-1.75}}{32} + \dfrac{7(2x)^{-2.75}}{128} + \dots$ 2. $X^{1/3} - x^{-2/3} - x^{-5/3} - \dfrac{5}{3} x^{-8/3} + \dots$ 3. $81x^4 - 216x^3 + 216x^2 - 96x + 16$ 4. $X^{1/5} - \dfrac{2}{5}x^{-4/5} - \dfrac{8}{25}x^{-9/5} - \dfrac{48}{125} x^{-14/5} + \dots$ 5. $243x^5 - 405x^4 + 270x^3 - 90x^2 + 15x + 1$ 6. $X^5 - 20x^4 + 160x^3 - 640x^2 + 1280x - 1024$ Ex. 6.5 1. a) x , y= -3.517, 1.877 (max) x, y=1.517, -61.877 (min) x, y = -1 , - 30 b) x, y= -1.608, 0.528 (max) ; x, y = -0.726, -0.158 (min) x, y = -1.167, 0.186 c) x, y= -3.43, 16.9 (max) ; x, y = 0.097, -5.05 (min) x, y = -1.67 , 5.96 d) x, y= -1.667, 50.81 (max) ; x, y = 3, 0 (min) x, y = 0.667, 25.4 e) x, y= -2.577 ; 0.385 (max) x, y = -1.423 , - 0.385 (min) x, y= -2 ; 0 2. a) -7 d) -14.25 b) -2 e) 2 c) -8 Ex. 6.6 1. ∞ 4. 1 2. -2 5. 0 3. 1 6. 0 Ex 7.1 1. a) -cosec x + c g) 2.94sin34x + c b) 56In x + c h) $3^{15x}\dfrac{1}{15\ln 3}$ + c c) 0.375 x^8 i) $-\dfrac{\cos 32x}{32}$ + c d) 2.4x^4 + 5.33x^3 + x^2 + 15x + c j) $\dfrac{1}{15}$ e^{15x} + c e) $\dfrac{\sin 32x}{32}$ + c k) 17x − 1.5x^2 + c f) x + c l) $\dfrac{1}{2}$ x^2 − 10x + c 2. a) 195.833 sq. units i) 56 893.18 sq. units b) 18 sq. units j) 2 sq. units c) 126.02 sq. units k) 2.67 sq. units d) 4.27 sq. units l) 0.5 sq. units e) 4.503 sq. units m) 258.3 sq. units f) 10.126 sq. units n) 0.365 sq. units g) 0.5787 sq. units o) 49.374 sq. units h) 25 775.34 sq. units

Don't miss out!

Visit the website below and you can sign up to receive emails whenever Efetobo Emede publishes a new book. There's no charge and no obligation.

https://books2read.com/r/B-A-OIGCB-DVNSC

Connecting independent readers to independent writers.

Also by Efetobo Emede

FET College Nated
Mathematics N4

About the Author

The author of this book has a Bachelor of Mechanical Engineering from the University of Benin, Benin City, Nigeria.

He has been lecturing since 2011 in the Private FET Colleges Nated in Johannesburg Cbd. He has been teaching the following subjects.

Mechanical Engineering N1-N6

Electrical Engineering N1-N6

About the Publisher

Mr. Efetobo Emede obtained a Mechanical Engineering Degree from the University of Benin, Benin City, Nigeria in February,1998.

He has been lecturing Mechanical Engineering N1-N6 and Electrical Engineering N1-N6 subjects since 2011.

The South African FET Colleges Nated Engineering qualifications are designed for prospective engineering diploma students.

www.ingramcontent.com/pod-product-compliance
Lightning Source LLC
Chambersburg PA
CBHW021156160726
47994CB00001B/244